Space and Society

Series editor

Douglas A. Vakoch, SETI Institute, Mountain View, CA, USA
and California Institute of Integral Studies, San Francisco, CA, USA

More information about this series at http://www.springer.com/series/11929

William Sims Bainbridge

The Meaning and Value of Spaceflight

Public Perceptions

Springer

William Sims Bainbridge
Arlington, VA
USA

ISSN 2199-3882 ISSN 2199-3890 (electronic)
ISBN 978-3-319-07877-9 ISBN 978-3-319-07878-6 (eBook)
DOI 10.1007/978-3-319-07878-6

Library of Congress Control Number: 2014945347

Springer Cham Heidelberg New York Dordrecht London

Cover illustration: by Paul Duffield

Printed on acid-free paper

Springer is part of Springer Science+Business Media (www.springer.com)

Preface

The conquest of outer space is in suspension, whether temporarily or permanently we cannot be sure. Great accomplishments of the past were the Apollo missions to the Moon, 1968–1972, and the first space probes to the planets, Mariner 2 to Venus in 1962 and Mariner 4 to Mars in 1964. Even the marvelous Hubble Space Telescope dates from more than two decades ago, 1990 to be exact. Human activity continues in near Earth orbit, although its value is open to debate, and astronomy continues to progress through use of space probes and space telescopes. While we cannot predict the future in space, and it seems quite problematic at the present time, research on the values of spaceflight for human beings can inform the important decisions that must be made, and illuminate the position of humans in the universe. This book draws upon a huge corpus of American public opinion data, and similar social science information, to explore the multiple meanings that exploration beyond the boundaries of our world may have.

The first chapter introduces the main methodologies and theories that must be employed to extract valid meaning from questionnaire data, using a few specific polls as illustrations. Two very different questionnaire methods must be combined: (1) administration of a few simple questions to random samples of the general population, to extrapolate with some confidence the balance of opinions in the society as a whole and (2) administration of much more complex questionnaires to specialized populations, placing the methodological emphasis on statistical analysis of how ideas fit together, using formal theory and empirical replication as validity checks. Two specific social-scientific theories are introduced that will feature throughout the book: (1) the standard observation that some individuals serve as opinion leaders, shaping the beliefs and attitudes of the general public and (2) technological determinism that analyzes any particular kind of technology in the context of the more general status of science and engineering of the particular historical period.

The next three chapters survey the development of public opinion using three different kinds of questionnaire study: (1) ordinary episodic public opinion polls like Gallup and Harris, (2) the General Social Survey (GSS) that systematically polled the US public for four decades, and (3) a specialized study of students at

Harvard University that explored their opinions about spaceflight more deeply than the two other approaches could afford to attempt. Chapter 2 focuses on the great Space Race between the United States and the Soviet Union, roughly in the decade and a half during 1957–1972, which is the period in which public opinion polls for the first time asked many questions about spaceflight, and popular awareness consolidated. Chapter 3 considers the period 1972–2012, using data from the GSS to see how support for space program funding correlated with support for other government programs and with variables describing respondents' age, social class, occupation, education, and political ideology. Chapter 4 employs data collected by the author at Harvard University in 1986, in the wake of the Challenger space shuttle disaster, through a pair of questionnaires that asked about a very large number of possible meanings spaceflight might have, employing the factor analysis statistical technique to identify underlying values, and determining how each affects overall support for space program funding.

The next two chapters employ the full range of kinds of questionnaire studies to place spaceflight in the context of world events and scientific progress. Chapter 5 returns to general public opinion polls to examine the meaning of events that took place after Apollo, especially policy decisions about the Strategic Defense Initiative and the recovery from the Challenger disaster, or might take place sometime in the future, notably the possible human return to the Moon and expeditions to Mars. Chapter 6 considers how spaceflight relates to various perspectives on science, beginning with a poll of scientists carried out in 1964 that found them rather unenthusiastic about the space program. This observation leads to the question of how science should be defined, whether as technical studies intended to provide information engineers can use to develop new technologies, or as philosophical explorations of the nature of reality as it really is, not as humans might wish it to be. Among the aspects of American culture that shape public perceptions of science, quite apart from factual news about space accomplishments are religion and pseudoscience, which do appear to militate against realistic appraisal, at least for significant minorities of citizens.

Three chapters then use questionnaires and comparable research techniques that have been developed recently to explore the popular culture of spaceflight, called *science fiction* or *sci-fi*. Chapter 7 examines the emergence of spaceflight fiction late in the nineteenth century, the launch of the first science fiction magazine in 1926 that established the genre, and the complex multidimensional set of genres that had consolidated half a century later, each with its own distinctive appraisal of spaceflight. Chapter 8 examines two more popular media, movies and television, given that cinema began depicting spaceflight as early as 1902, and a very significant number of films and programs continued to do so, especially after about 1950, using recommender system data on recent movie preferences to identify multiple mass media conceptions of interplanetary travel. Chapter 9 considers the newest mass medium that depicts spaceflight, computer games, especially massively multiplayer online (MMO) virtual worlds, in which users experience simulated spaceflight, including questionnaire-like data from two of these MMOs that suggest the human goals that the respondents seek beyond the boundaries of the Earth.

The concluding chapter looks back at all the findings of earlier chapters in the context of general explanatory theories. Its starting point is the frontier metaphor repeatedly associated with space exploration, especially in the light of the theory of what happens when a frontier closes, enunciated over a century ago by American historian Frederick Jackson Turner. A larger context can be provided by several theories, primarily European in origin, about the fall of civilizations, that would consider the end of space exploration to have dire consequences for humanity. The chapter then considers how the spaceflight social movement competes with other cultural traditions within western societies, giving some attention to the links between spaceflight support and gender, and with education analyzed by gender. Some questionnaire data suggest that the worldwide explosion in popular use of the Internet may be creating a new world culture that is more favorably disposed toward space exploration. Technological determinist theories suggest that spaceflight may experience a second acceleration phase, so long as popular interest has some degree of strength, once other fields of technology advance to the point at which new means of interplanetary travel become possible.

Contents

Chapter 1
Background

Over a brief period of time, within the entire context of human history, men have visited the Moon and robot spacecraft have explored the full width of the solar system. Yet the future of spaceflight is uncertain, and plans for an expedition to Mars dating back as much as seven decades are far from realization. Both to identify the basis for possible future progress and to understand one of the great human challenges, we need to know what spaceflight means. It may mean many things, serving different values for different people, so research on the meaning of spaceflight can become a tool for understanding human meaning in general.

This is a book of history, charting and analyzing public opinion, political ideology, and artistic expression related to space exploration, using data from 1938 until 2012. Included are many standard public opinion polls, not merely reporting responses to individual questions but often analyzing raw data from Harris Polls and the General Social Survey, to identify factors that shape perceptions of space. Other questionnaire surveys of special populations, often using far more questions about space than are found in any ordinary poll, were carried out by the author. In addition, the proliferation of new online forms of data analogous to opinion polls enable new forms of analysis, some of which have never been tried before.

Unfortunately, social science has tended to ignore spaceflight as a research topic, leaving the field to advocates and historians, so one function of this book is to bring together information from many sources that have not previously been considered as a whole in the light of theory. Many potential insights will be offered, more as hypotheses than as confident findings, in hopes that social scientists will begin serious work on issues crucially important for the future of humanity. Only after each idea has been debated in the journals, and tested through replication with new datasets, can we be sure what the real meaning of spaceflight is, and how it illuminates the shape of things to come.

W.S. Bainbridge, *The Meaning and Value of Spaceflight*,
Space and Society, DOI 10.1007/978-3-319-07878-6_1

1.1 A Time of Indecision

On July 21, 2011, during the 42nd anniversary of the Apollo 11 flight to the Moon, the landing of Atlantis ended the space shuttle program. Historian Roger Launius has called the shuttle, "a vehicle filled with contradictions and inconsistencies" (Launius 2006). While we cannot know how American spaceflight would have developed without the shuttle, it had over-optimistically promised to render spaceflight cheap and safe. The fatal disasters that ended the last flights of Challenger and Columbia contradicted the hope of safety, and the failure of the United States to develop a second-generation shuttle reflected the economic realities. The Bush and Obama administrations launched their own new spacecraft, using the words *launch* and *new* only in a figurative sense, because neither the Constellation nor the Space Launch System had left the Earth by the time this book was published, and both seemed to be remakes of the antique Saturn rockets that first flew half a century before the last flight of Atlantis.

What did this final shuttle flight mean? Reporting results from a public opinion poll administered during the last Atlantis mission, *The Christian Science Monitor* sought to answer this question about the space shuttle program: "Were Americans sad to see it go?" Of the 904 people who were asked, 56 % opposed cancellation of the shuttle program, and 52 % felt it had justified the cost. Yet when asked what should happen with NASA funding, only 10 % wanted it increased. A plurality of 49 % were happy to see it stay the same; 28 % wanted it decreased, and 8 % wanted to end NASA (Sappenfield 2011). Using data from the same poll, *Investor's Business Daily* reported that only 9 % of respondents believed that "the current administration has a clear plan for space exploration," 18 % were not sure, and fully 72 % were convinced it did not (Merline 2011).

A CNN poll of 1,009 Americans, carried out during the flight and released at the time of the landing, focused on the future, asking about the national impact of the termination of the shuttle and the prospects for a successor. It must be admitted that the poll's introduction may have biased the pattern of responses: "As you may know, the current space shuttle mission will be the final time that the U.S. will send astronauts into space using the shuttle. Until the U.S. develops a replacement for the shuttle, all manned U.S. space flights will take place in spacecraft that are owned by other countries. Overall, do you think the end of the space shuttle program will be good for the U.S., bad for the U.S., or not have any effect on the U.S. at all?" Just 16 % were willing to resist this suggestion that cancellation would be bad, calling it good, 33 % said it would have no effect, and 50 % gave the expected response that this situation would be bad. In response to other questions, fully 87 % predicted the US will develop "a replacement spacecraft that will be capable of sending U.S. astronauts into space and returning them to Earth," and 75 % said the US should indeed do so (CNN/ORC Poll press release 21 July 2011).

The highly respected Pew Research Center conducted its own poll of 1,502 adult residents of the US a month before the last Atlantis flight, using somewhat better methodologies of data collection and analysis to explore the meaning of the shuttle program (Pew Research Center 2011). The first space-related question in the Pew poll asked, "Do you think the space shuttle program has been a good investment for this country, or don't you think so?" While 9 % declined to answer, 55 % called it a good investment. In its report of results, Pew contrasted this fraction with the 70 % who called it good investment in June 1986, after the January 1986 Challenger disaster, and 66 % who did so in August 1981, 4 months after the first shuttle orbital mission. It is noteworthy that the earlier polls used slightly different wording, "is a good investment" rather than "has been a good investment," reflecting the fact that the shuttle had a future during the earlier years.

The Pew survey also asked several questions about the respondent, for example finding that 59 % of men thought it was a good investment, versus 52 % of women. Republicans were more supportive than Democrats, 63 % to 48 %, and there was a similar difference between those with annual family income more than $75,000 versus less than $30,000, 67 % compared with 44 %. The difference by education was about the same, 66 % of colleague graduates calling the shuttle a good investment, compared with 47 % of those with high school educations or less.

Four questions sought to learn what values the space program served for many Americans. One asked, "In your view, is it essential or not essential that the United States continue to be a world leader in space exploration?" While 4 % expressed no opinion, 58 % felt it was essential, and 38 % felt it was not. The three other value questions began: "Thinking about the space program more generally, how much does the U.S. space program contribute to…" Here are the percents who answered "a lot:"

38 % Scientific advances that all Americans can use
34 % This country's national pride and patriotism
39 % Encouraging people's interest in science and technology.

During the years around the end of the shuttle program, the Gallup Poll measured public support for the space program more generally, for example in July 2009 finding that 58 % of Americans felt the space program had been worth the investments in it. One demographic variable influencing attitudes was age, and the Gallup report remarked, "Notably, those old enough to remember the historic moon landing are actually somewhat less likely than those who are younger to think the space program's costs are justified. Among Americans aged 50 and older (who were at least 10 years old when the moon landing occurred), 54 % think the space program's benefits justify its costs, compared with 63 % of those aged 18–49" (Jones 2009). In later chapters we shall consider both changing public opinions about space exploration over six decades of history, and the influence of age.

In May 2013, Gallup asked poll respondents to rate the performance of nine US federal government agencies, including NASA, finding that 42 % rated "the job being done by" the space agency excellent or good, and an identical 42 % rated it fair or poor (Jones and Saad 2009). Gallup calculated a summary "net positive" rating by subtracting the poor responses from the combined excellent-good responses, which in this case was $42 - 10 = 32$. This was way below the net positive rating for the Centers for Disease Control, which was 52, and a significant drop from NASA's net positive rating back in 2009, which had been 51. Net positive ratings of six of the nine agencies had dropped from 2009 to 2013, perhaps reflecting increased doubts about the competence of government in general, but the NASA decline could reflect the impact on public opinion of the end of the shuttle program in the absence of much publicity about the Space Launch System or what its real goal might be.

When opinion poll data are reported in the journals, books, or websites devoted to spaceflight, they tend to be seen as a resource for understanding how to increase public support and thus public funding for NASA. Social scientists unconnected to the space program almost never look at these data, let alone analyze them in the contexts of theory and of data about other variables. That will be our job here. The best of the past space scholars who wrote about public opinion data, notably Roger Launius but also Alan Steinberg and Wendy Cobb, published in the journal *Space Policy*, thereby providing expert advice for people who wanted to build the future human presence in space (Launius 2003; Steinberg 2011; Cobb 2011). They debated the impact of events on public opinion, and the characteristics of people who were more or less enthusiastic, but they generally reported that the public did not understand the realities of NASA funding, let alone the technical details of spaceflight itself.

It would be wrong to conclude that public opinion does not matter, because at the very least a complete loss of public support could lead politicians to downgrade NASA even further in their funding priorities, while a surge of interest coming at a decision point about some new project could tip the balance in favor of funding it. For our purposes here, public opinion data are the gateway to deeper understanding of the meaning of spaceflight, especially as we explore the data rather more intensively than has been done before, and begin to investigate the thoughts not only of random samples of the general population, but of groups and subcultures that may understand spaceflight better than the average. We shall consider vast troves of questionnaire data in this book, along with other kinds of data that are similar in the ways they can be analyzed statistically and understood conceptually. Prior to entering that treasure house of information, we will need several kinds of scientific orientation, beginning with an overview of the history and methodology of opinion polls, which underwent the most rapid progress in the 1930s.

1.2 The First Spaceflight Opinion Poll

Arguably the very first American opinion poll about spaceflight was conducted immediately after Halloween in 1938. It did not concern Robert Goddard's ideas about liquid fuel rockets, nor the early discussions of building a 200-inch reflecting telescope on Mount Palomar, but the Orson Welles radio dramatization of *The War of the Worlds* by H. G. Wells. In style, the program was a series of increasingly frantic news reports about a Martian invasion in New Jersey, and apparently people who tuned in late were unaware it was a drama but thought it reported real events. A Gallup Poll indicated that as many as 1,700,000 listeners believed this, and 1,200,000 were frightened by it. A questionnaire administered to school principals suggested that perhaps a quarter million children were frightened. News media widely reported frantic behavior in their areas (Cantril 1940, 1941). This may seem bizarre and substantively unimportant, but the incident is significant for three reasons.

First, it seemed to demonstrate the power of the mass media to shape public opinion. At the time, radio was a growing industry, and opinion polls were also new, so the results of this poll supported the economic interests of both broadcasters and pollsters by apparently demonstrating that both were important. However, reports of mass panic may have resulted from both errors and media hype, and no real panic may have occurred. A chance to test this possibility came in 1973, when Swedish radio broadcast a fictional news report indicating that a nuclear power plant had just exploded and was spreading lethal radiation all over a wide area. News reports of panic were triggered by the fact that a few citizens reasonably enough called the local police or news outlets asking what was happening. One radio station leapt to the erroneous conclusion there was a panic, and other news agencies picked up that false story. A scientific poll determined that only 2 % of the population had taken any action, under the impression the news was correct, such as shutting their windows (Rosengren et al. 1975).

Second, the Martian invasion poll data indicated that some people worried by the radio program had very reasonably sought further information, for example calling the police, or simply changing to a different radio station to see if it also was reporting an invasion. This may seem obvious, but findings like this led over the following decade to the emergence of a sophisticated two-stage theory of the spread of information through the general public (Katz and Lazarsfeld 1955; Pooley and Socolow 1972). First, local opinion leaders who are attentive to the news collate information in their own minds, often better educated than the average. Then, these opinion leaders influence others in the community, who are less inclined to develop their own interpretations of the news. This perspective will greatly inform this book, as we shall often consider the views of people who may be opinion leaders, without entirely disregarding the views of the majority. With specific reference to spaceflight, opinion leaders in government and in social movements have been especially influential, and it is an open question whether any of the steps in the development of space technology were shaped by the opinions of the general public. Given that our theme is the meaning of spaceflight for human beings, the views of the public are still important, but as indicators of a variety of meanings.

Third, the Martian invasion episode suggests two ways in which spaceflight may have meaning. First, it was broadcast at the end of October 1938, half a year after Germany had annexed Austria, and just days after Germany took the Sudetenland from Czechoslovakia. War worries were rampant, so the invading Martians became symbols for Germans, or for military dangers more generally. Often in subsequent years spaceflight may have taken on meaning as a symbol for something else. Second, the Martian invasion was a fantasy, and future human colonization of Mars may also be a fantasy. To feel that Martians might exist required one to believe Mars was more similar to the Earth than the robot orbiters and landers sent there discovered, beginning with Mariner 4 in 1964. Another way to express this is that spaceflight may become a metaphoric vehicle for humanity's fears and hopes, both of which may be misplaced.

During the Second World War, the American public was certainly not well informed about the rocket technology developments taking place in Germany. In July 1944, 2 months before the first military use of the V-2 rocket, the Gallup poll asked a pair of rather ambiguous questions, ultimately triggered by rumors based on fact: "A Swedish newspaperman says the Germans are now building robot bombs which can hit cities on our East Coast. Do you believe this is true? Do you think that in another 25 years such flying bombs will be able to cross the Atlantic Ocean?" (Gallup 1972). It is left unspecified whether these "robot bombs" or "flying bombs" were V-2 rockets or V-1 unmanned pulse-jet aircraft, but the rumor was probably based on V-2 tests carried out over the North Sea from the German rocket development base at Peenemünde. Neither machine had the range to reach the United States, but either technology could have been developed to do so. Just 20 % of American respondents to the Gallup Poll felt the Germans were already developing such long-range robot weapons, but 70 % believed they could be developed before 1970, which proved to be correct.

Gallup asked a question specifically about spaceflight in October 1947: "How long do you think it will be before man will be able to fly to the moon?" Of those who selected a specific range of years, the median chose 20–29 years. However, 16 % failed to respond to the question, 23 % said they could not guess, and the largest group, 38 %, answered "never." In December 1949, Gallup asked the question a different way: "In the next 50 years, do you think men in rockets will be able to reach the moon?" Just 15 % of respondents confidently answered "yes," a fraction that had increased to 38 % when the question was asked again in January 1955 (Gallup 1955). Clearly, awareness of the real possibility of space travel was growing, but extensive public opinion polling on the subject did not really begin until the launch of Sputnik 1 in 1957, as we shall chart in the second chapter of this book.

1.3 Principles of Public Opinion Polling

As rocketry was developing from the first liquid-fuel launch by Goddard in 1926, public opinion polling was also developing. In business and government, the equivalents of questionnaires have existed for centuries, and by the middle of the nineteenth century the US Census was collecting rather detailed information about each household, through questionnaire-based interviews conducted by enumerators who went door-to-door. Arguably, social-science was the leader in the development of electronic computing, when Herman Hollerith developed technology for analyzing data from the 1890 US Census and founded The Tabulating Machine Company in 1896, a precursor of IBM (Bainbridge 2004).

A census seeks information about every individual person within its geographic scope, and is exceedingly expensive to do with large populations. Therefore, public opinion polls must find a way of describing the entire population on the basis of a representative sample, and issues of sampling have posed problems for the field throughout its history. The most famous example is the debacle associated with the 1936 US presidential election that discredited simplistic polling methods.

In the elections of 1920, 1924, 1928 and 1932, a popular magazine called *The Literary Digest* had correctly predicted the winner, through commercial polling methods, using for example telephone directories to identify people to whom paper questionnaire should be mailed. By October 31, 1936, ten million questionnaires had been sent out, and 2,376,523 had been returned. Before we consider the results, four things should be noted. First, these are huge numbers, far larger than covered by the many other polls reported in this book. Second, the poll was really the equivalent of a sophisticated advertising campaign for the magazine and its clients, thus affordable on grounds other than obtaining scientifically valid data. Third, less than a quarter of the people who received a questionnaire filled it out and sent it back, a severe example of non-response bias, the likelihood that those who answered are atypical in motivations and thus attitudes. Fourth, 1936 was the middle of the Great Depression, which disrupted many people's lives, thus rendering obsolete some of the lists used to draw the sample of respondents, and rendering problematic many assumptions about the coherence of the American body politic.

The 1936 Literary Digest Poll confidently predicted that president Franklin D. Roosevelt would be thrown out of office, but in fact he was re-elected. This proved to be a marvelous advertising coup for George Gallup, who had founded his polling firm the year before. Gallup had sent 3,000 postcard ballots to a random sample of people on the lists of *The Literary Digest*, getting the same result the magazine did. But he also sent 3,000 postcard ballots to his own sample of voters, correctly predicting the outcome of the election (Gallup 1976).

From Gallup's day until now, conventional public opinion polls have actually combined methods in various ways, balancing cost against representativeness against the kinds of information that can be obtained. Gallup used combinations of *random* and *quota* sampling. To do a proper random sample, one needs to have essentially a list of everybody in the population, and use some random number system to identify the individuals who will be polled. In the case of voters, such lists exist, but may be out of date, and one cannot expect all the people who receive questionnaires to answer the questions and send them back. For example, Chap. 4 will report on a pilot study of voters in the Seattle, Washington, area, in which the response rate was 45 %, and assess whether that was sufficient given the goals of the study.

Quota sampling attempts to compensate for the fact that perfect random samples cannot really be obtained, by making sure that the sample has the same distribution as the population on some key variables, that are believed to affect the estimates of the opinions in question. For example, when Barak Obama was running for the presidency, it would have seemed important to make sure the respondents to a poll included the same fraction of members of minority groups who belonged to the electorate. A kind of geographic quota system may be necessary in presidential election polls, if the goal is to predict the outcome of the electoral college as well as the popular vote, because the votes of people in low-population states count more than those in high-population states. An alternative to quota sampling is to weight responses by different groups in the sample, using statistical procedures, although this weighting can be especially inaccurate when a group is heavily underrepresented in the sample.

This book will find merit in questionnaire data that were obtained following a wide range of methodologies, even in the later chapters in data that only resemble questionnaire data. One reason is that the public opinion model of questionnaire research is not the only one that has proven valuable for social science. Another model, primarily developed in personality and social psychology, does not attempt to predict elections or describe the average views of the general public, but to seek and test theories about the alternative sets of values and conceptualizations held by diverse groups of people.

The two approaches can also be distinguished in terms of the kinds of questions they ask. Public opinion polls ask very simple questions, which almost anyone can understand and might have an opinion about, usually a very small number of questions about any topic that are rather superficial. Social psychological questionnaires typically use batteries of many items in the same general area, investing its statistical analysis not in sampling issues, but in exploring how these many items fit together in respondents' minds. Validity is achieved not through the sampling procedures, but through careful examination of how the results compare with the theories of interest to the scientists, and the ability of the results to be replicated in other studies that use different sets of respondents.

This book brings together a vast amount and diversity of data about people's perceptions of spaceflight, and uses a diversity of methods to make some sense of it. The chief challenge, and the opportunity that made this book both possible and necessary, is the fact that social science has ignored this topic, so we lack a well-developed scientific literature about the meaning of spaceflight to human beings. Many of the results reported here may seem obvious once they have been stated, and many of them may be found scattered across existing reports about single opinion polls. Yet it is useful to bring all these insights together, and compare information from many sources. But we shall also derive many fresh hypotheses by interpreting particular patterns or anomalies that arise in this comprehensive analysis.

Intentionally, this book avoids very complex statistical analysis of the kind that might be needed to test hypotheses and resolve scholarly disagreements. The reason is not merely that the intended readership is broad and few readers would possess the necessary training in statistical analysis. Rather, intensive quantitative analysis is best presented through scientific journals, and subjected to debate among many competent social scientists. Thus, each of the arguments in favor of an hypothesis presented here is intended to render the hypothesis clear and plausible, but not in most cases to prove whether it is true. My own view is that most cogent hypotheses about human attitudes are true in at least a few cases, and the real problem is to weigh their relative strength while a diversity of social forces play out in our culture. That requires a scientific community devoted to exploring the human meaning of spaceflight in the context of more general research and debates about the human future. This book is a step toward creation of that community.

1.4 The Final Frontier

Many different publics hold a variety of views on the meaning of spaceflight, and social science itself does not speak with one voice. Writing in the *International Encyclopedia of Social Science*, I pointed out that the facts of the history of space exploration to date are clear, on the basis of a huge library of technical and scholarly publications, but the social-scientific interpretation is hotly debated (Bainbridge 2008). The view around 1960 was that international propaganda competition was the main driver, as has been summarized by Vernon van Dyke (van Dyke 1964). Sociologist Amitai Etzioni argued in 1965 that the American space program was a useless extravagance through which the military-industrial complex looted the national treasury (Etzioni 1965). Then in 1970, John Logsdon argued that President John F. Kennedy's decision to go to the moon was a means for reviving the political spirit of his New Frontier program after foreign policy defeats in 1961 with the aborted Bay of Pigs invasion of Cuba and being browbeaten by Soviet leader Nikita Khrushchev in a summit meeting (Logsdon 1970).

In my 1975 sociology doctoral dissertation and first book, I suggested a very different analysis that placed world politics in a secondary role: In Germany and the Soviet Union, as well as in the United States, leaders of the transcendental spaceflight movement had cleverly manipulated beleaguered political leaders to invest in space as a symbolic solution to their inferiority in competition with other leaders (Bainbridge 1976). Michael Neufeld has argued against this thesis in the case of Germany, asserting that technically competent military engineers possessed a correct estimation of the military potential of the technology (Neufeld 1995). Walter McDougall argued against this view in the case of the Soviet Union, stating that Marxist ideology naturally supported visionary technological projects (McDougall 1985). More recently, John Logsdon has argued that the American space program has been trapped in a vicious circle, as members of the movement convince political leaders to undertake technically demanding projects, but the public is not willing to invest enough to make them successful (Logsdon 2006).

Clearly, human spaceflight is in something like a holding pattern, and to a great extent has been since the last Apollo mission to the Moon over four decades ago. Two kinds of project have served to fill the gap, orbiting manned space stations and the space shuttle, but the original concept of "space station" would be orbiting platforms from which to launch human missions far from Earth, and the space shuttle took its last flight in 2011. Unmanned spaceflight has shown the capacity for constant progress, with ever more sophisticated deep space probes, landers, and Earth-orbit telescopes. A key question, therefore, will be whether the public can shift its interest from human space adventures to scientific research carried out by means of machines. However, a classic theory in social science called *technological determinism* discounts both public opinion and the behavior of societal elites, and thus would analyze space development as the reflection of more general technological progress.

While small, solid-fuel military rockets had existed for centuries, rockets capable of achieving orbit could not have been built before the twentieth century. The kind most suitable for spaceflight, liquid-fuel rockets employing high energy propellants, required the development of technologies for super-cooling at least oxygen and perhaps hydrogen as well to store these gasses as liquids in the fuel tanks. Many launch vehicles have used liquid oxygen as the oxidizer, combining in the combustion chamber with a more mundane fuel such as alcohol, gasoline, or kerosene. Storable but less energetic liquid propellants, such as hydrazine, have applications for spaceflight, such as in thrusters to control the orientation of a spacecraft, or to land on a distant moon or planet. But like gasoline, hydrazine required the development of industrial chemistry in the nineteenth century, and was not available much before the twentieth century.

Historians and social scientists with the technological determinist perspective often study the emergence of specific technologies, but do so in the much broader context of all technologies that had been developed prior to the particular point in time. The classic example is that the rise of cities could not occur until after the development of systematic agriculture. This occurred at what V. Gordon Childe called the *Neolithic Revolution*, in which the term *Neolithic* refers to the new stone age in which stone tools had developed considerable sophistication and diversity (Childe 1951). Farming required not only tools, storage facilities, and skill in using natural resources to construct them, but also domestication of plants and animals. As a complex socio-technical system developed during the Neolithic Revolution, human population began to increase, specialization in skills and tools initiated the division of labor in which individuals began to perform distinguishable jobs, and increasingly complex political and religious institutions emerged to manage the growing societies. As villages evolved into cities, entirely new forms of technical and social systems were required to sustain them.

Technological determinists often wrote essays seeing to refute common notions about human progress, which they criticized as overly romantic. For example, Leslie White dismissed the significance of the Dark Ages after the fall of the Roman Empire, asserting that technological progress continued during that period despite the decline of elite culture (White 1959). Robert K. Merton minimized the importance individual inventors, noting that new ideas were typically invented many times, and separate invention only ceased when the innovation became widely known (Merton 1973). S. C. Gilfillan put the point thus: "There is no indication that any individual's genius has been necessary to any invention that has had any importance. To the historian and social scientist the progress of invention appears impersonal" (Gilfillan 1963).

Gilfillan's views are especially relevant here, because much of his own research on the history of technology concerned sailing ships, which are an obvious metaphor for spaceships. Even before the introduction of steam engines early in the nineteenth century, ocean-going vessels were extremely complex, requiring application of a variety of manufactured materials and crew skills. Another metaphor for space travel is the railroad, which was the focus of an edited volume of essays by many authors, titled *The Railroad and the Space Program: An Exploration in Historical Analogy* (Mazlish 1965). Although that book chiefly looked at the socio-economic impact of the railroad as a source of insights about the impact of the space program, both steamships and railway trains illustrate the fact that radically new transportation technologies require a very large number of prior technical developments. Thus, spaceflight is not really a single invention, but a bundling together of many existing inventions, adding just a few new ones to achieve a new goal.

Some technological determinists did not dismiss human initiative altogether in their theories of how technology developed, but did place it in a subordinate position to other factors. The clearest example is the theory of social change published way back in 1922 by William F. Ogburn, that views human history as the result of a complex interaction among four discernable processes (Ogburn 1922):

1. Invention: A new technical innovation emerges, not because one inventor has a brilliant idea, but because society reaches the point at which the knowledge and other factors required for the invention have collected.
2. Accumulation: The general stock of technological capabilities grows, because new things are invented more rapidly than old ones are forgotten.
3. Diffusion: Innovative ideas spread from one cultural group to another, given that groups may invent in different areas, depending upon accidents of history and natural resources.
4. Adjustment: Non-technical aspects of a culture respond to invention, sometimes with difficulty because new social institutions are required, and old ones may become obsolete.

Note that diffusion feeds into accumulation, and both increase the basis for invention. Ogburn said that social movements can play noteworthy roles, but chiefly in the society's adjustment to new innovations. He specifically referred to *cultural lag*, a maladjustment that comes about because the various parts of culture are not changing at the same rate. Rapid progress in one area may demand progress in another area related to it, but the adjustment is delayed, perhaps for many years. It may be that public ignorance about or indifference to spaceflight is an example of cultural lag, which could be overcome simply by the passage of time, or the educational efforts of the spaceflight social movement. Social scientists in the technological determinist tradition often write about cultural lag, but they less often consider that if technology determines itself, then some conceivable development cannot occur until certain other developments have already been achieved.

Perhaps, therefore, the current stasis we see in spaceflight development has nothing to do with public indifference or the ignorance of political leaders. It may instead reflect the need to wait for other fields of technology to advance until they could enable a new wave of astronautical innovation. The most obvious example is controversial, but clear enough in its technical aspects to make it worth mentioning here: Nuclear propulsion during launch from Earth to orbit (Bussard and DeLauer 1965; Gross 1970). It is one thing to use modest radioisotope thermoelectric generators to produce the electric power required by the two Voyager probes that were launched by ordinary chemical rockets back in 1977 to explore the outer solar system, and quite another to use high-power nuclear rockets to launch heavy payloads, with all the hazard to the environment that would pose (Maharik and Fischoff 1993). Rocket engines based on nuclear fission were developed but abandoned over 40 years ago, and controlled nuclear fusion has defied attempts to develop it for any purpose. But if some new and vastly more efficient means were developed to lift payloads into orbit, a new Age of Space could dawn (Coopersmith 2011).

The concluding chapter of this book will reconsider the full range of relevant social-scientific theories, in the light of the empirical findings of all the other chapters, but a few points deserve quick mention here. After Ogburn, sociologists tended to focus primarily upon diffusion of innovations, just one of his four points. Economists tended to focus on none of the four, because they assumed innovation would continue so long as free markets motivated entrepreneurs to innovate, or perhaps the accumulation of capital was a subprocess within Ogburn's accumulation concept. Technological determinism was brought into doubt by environmentalists and social scientists from many fields who happened to be concerned about sustainability, but they almost invariably thought within the confines of a limited Earth, rather than imagining that colonization of other planets could transcend all resource limitations. Environmentalists hold a wide range of views, but one is that we should transition from wrongly named technological "progress" to appropriate technology, which may not change over the centuries after the world stabilizes in a sustainable system (Schumacher 1973). Visionary advocates of space exploration may also hold many divergent views about the future of society here on Earth, but continued technological development is required to expand humanity throughout the galaxy.

References

Bainbridge, W. S. (1976). *The spaceflight revolution*. New York: Wiley-Interscience.

Bainbridge, W. S. (2004). Hollerith card. pp 326–327. In: Bainbridge WS (ed.), *Berkshire Encyclopedia of Human-Computer Interaction* (pp 326–327) (Great Barrington, Massachusetts: Berkshire, 2004).

Bainbridge, W. S. (2008). Space exploration. In W. A. Darity Jr (Ed.), *International encyclopedia of the social sciences*. Macmillan Reference USA: Detroit.

Bussard, R. W., & DeLauer, R. D. (1965). *Fundamentals of nuclear flight*. New York: McGraw-Hill.

Cantril, H. (1940). *The invasion from Mars: A study in the psychology of panic* (pp. 55–63). Princeton, New Jersey: Princeton University Press.
Cantril, H. (1941). *The psychology of social movements*. New York: Wiley.
Childe, V. G. (1951). *Man makes himself*. New York: New American Library.
CNN/ORC Poll Press Release (21 July, 2011). Retrieved November 7, 2013, from i2.cdn.turner.com/cnn/2011/images/07/21/poll.july21.pdf.
Cobb, W. N. W. (2011). Who's supporting space activities? *Space Policy, 27*, 234–239.
Coopersmith, J. (2011). The cost of reaching orbit: Ground-based launch systems. *Space Policy, 27*, 77–80.
Etzioni, A. (1965). *The Moon-Doggle*. Garden City, New York: Doubleday.
Gallup, G. (1955). Number who think trip to Moon possible has doubled in 5 years, press release, January 28, American Institute of Public Opinion, Princeton, New Jersey.
Gallup, G. H. (Ed.). (1972). *The Gallup Poll: Public opinion 1935–1971* (p. 456). Random House: New York.
Gallup, G. (1976). *The sophisticated poll watcher's guide* (pp. 64–67). Ephrata, Pennsylvania: Science Press.
Gilfillan, S. C. (1963). *The sociology of invention* (p. 10). Cambridge, Massachusetts: MIT Press.
Gross, E. (1970). Problems of nuclear rockets. *Science News, 97*(18), 440–441.
Jones, J. M. (2009). Majority of Americans say space program costs justified, press release, July 17, Gallup. Retrieved November 9, 2013, from www.gallup.com/poll/121736/majority-americans-say-space-program-costs-justified.aspx.
Jones, J. M., Saad, L. (2009). Americans sour on IRS, rate CDC and FBI most positively, press release, Gallup. Retrieved November 9, 2013, from www.gallup.com/poll/162764/americans-views-irs-sharply-negative-2009.aspx.
Katz, E., & Lazarsfeld, P. F. (1955). *Personal influence: The part played by people in the flow of mass communications*. Glencoe, Illinois: Free Press.
Launius, R. D. (2003). Public opinion polls and perceptions of US human spaceflight. *Space Policy, 19*, 163–175.
Launius, R. D. (2006). Assessing the legacy of the Space Shuttle. *Space Policy, 22*, 226.
Logsdon, J. M. (1970). *The decision to go to the Moon*. Cambridge, Massachusetts: MIT Press.
Logsdon, J. M. (2006). 'A failure of national leadership:' Why no replacement for the Space Shuttle. In S. J. Dick & R. D. Launius (Eds.), *Critical issues in the history of spaceflight* (pp. 269–300). Washington, D.C.: National Aeronautics and Space Administration.
Maharik, M., & Fischoff, B. (1993). Public views of using nuclear energy sources in space missions. *Space Policy, 9*(2), 99–108.
Mazlish, B. (Ed.). (1965). *The railroad and the space program: An exploration in historical analogy*. Cambridge, Massachusetts: MIT Press.
McDougall, W. A. (1985). *The heavens and the Earth: A political history of the space age*. New York: Basic Books.
Merline J (2011) Majority opposes shuttle shutdown. *Investor's Business Daily*, July 20. Retrieved November 7, 2013, from news.investors.com/ibd-editorials-polls/072011-578923-majority-opposes-shuttle-shutdown.htm.
Merton, R. K. (1973). *The sociology of science: Theoretical and empirical investigations*. Chicago: University of Chicago Press.
Neufeld, M. (1995). *The rocket and the Reich: Peenemünde and the coming of the ballistic missile era*. Cambridge, Massachusetts: Harvard University Press.
Ogburn, W. F. (1922). *Social change with respect to culture and original nature*. New York: B. W. Huebsch.
Pew Research Center (2011) Majority sees U.S. leadership in space as essential, report, July 5. Retrieved November 9, 2013, from people-press.org/2011/07/05/majority-sees-u-s-leadership-in-space-as-essential/.
Pooley, J. D., & Socolow, M. J. (1972). Checking up on *The Invasion from Mars*: Hadley Cantril, Paul F. Lazarsfeld, and the making of a misremembered classic. *International Journal of Communication, 7*, 1920–1948.

Rosengren, K. E., Arvidson, P., & Sturesson, D. (1975). The Barsebäck 'panic:' A radio programme as a negative summary event. *Acta Sociologica, 18*, 147–162.
Sappenfield M (2011) Space shuttle program ends: Were Americans sad to see it go? *The Christian Science Monitor*, July 21. Retrieved November 7, 2013, from www.csmonitor.com/USA/2011/0721/Space-shuttle-program-ends-Were-Americans-sad-to-see-it-go.
Schumacher, E. F. (1973). *Small is beautiful: Economics as if people mattered.* New York: Harper and Row.
Steinberg, A. (2011). Space policy responsiveness: The relationship between public opinion and NASA funding. *Space Policy, 27*, 240–246.
van Dyke, V. (1964). *Pride and power: The rationale of the space program.* Urbana: University of Illinois Press.
White, L. A. (1959). *The evolution of culture*. New York: McGraw-Hill.

Chapter 2
The Space Race

The extraterrestrial competition between the Soviet Union and United States that began in the late 1950s gave prominence to some of the potential meanings of space exploration to the exclusion of others. By the 1960s, serious analyses had begun to appear in social science publications, and major polls like Gallup and Harris included items about the space program (Goodwin 1965). Initially, few citizens understood anything about spaceflight, or much about the solar system, so one trend reveals increasing awareness. Depending on exactly what questions were asked, citizens always showed great disagreement over what priority should be given to the American space program. Generally the majority was opposed to increased funding, but there was sufficient public support so that political elites could invest in the program. An additional facilitator of constant support was the usually non-partisan nature of the program and support for it, although competition with the Soviet Union was the broader political factor encouraging progress.

This chapter will raise many issues considered more deeply in later chapters, but it also continues the discussion of the role of leaders in setting the terms of public opinion. These leaders are of two kinds. Some are members of the general public, typically better educated than the average, better informed, and more influential in their local communities. In the jargon of sociology they are called *opinion leaders*, because they are leaders *within* public opinion. Others are prominent in social movements, political factions, and established interest groups. We can call them *organization leaders*, who influence public opinion from outside, being or seeking to become members of a societally influential elite class. Both concepts raise questions about the nature of democracy, to which we shall return in the concluding chapter.

During the Space Race, the general public was primarily a passive observer, organization leaders were dominant, and opinion leaders functioned as mediators. This certainly does not mean that public opinion polls are useless to understand the meaning of spaceflight during this time. Quite the opposite, because they help us to identify the messages that were being transmitted to the public, and to begin to understand how spaceflight connected to some values that various members of

W.S. Bainbridge, *The Meaning and Value of Spaceflight*,
Space and Society, DOI 10.1007/978-3-319-07878-6_2

the citizenry already held. Advocates for the space program tend to view polls as marketing advice, highlighting which of the potential meanings of spaceflight they should emphasize in their propaganda. But they can also provide insights about what spaceflight could really mean for human beings more generally, and thus potentially providing guidance about what kind of space program, devoted to what goals, we might want to create for the future.

2.1 Public Indifference

One remarkable aspect of this situation is the fact that there did not exist vocal foes of space exploration, whose public arguments dissuaded citizens from supporting it. This was not the case, of course, for many other public issues. The most extreme example was the Civil War, 1861–1865, which school children are taught was fought to end slavery, but which also involved issues of state's rights and competition between two fundamentally different economic systems. A more recent example was prohibition of alcoholic beverages, 1920–1933, which was promoted by the Temperance social movement, and opposed by liquor companies, taverns, and their customers (Gusfield 1963). During the period 1950–1970, three political issues were bipolar: the crusade to suppress Communists and comparable radicals inside the United States, school desegregation and the Civil Rights Movement, and the Vietnam War. Space exploration was not a bipolar issue, because the social movement promoting it lacked a clear political opponent.

Throughout this book we shall emphasize the many reasons why people might support space exploration, each of which identifies a different possible meaning it might have. Yet meanings can be found on the negative side of the public opinion ledger as well. Here are four possible explanations for public indifference:

1. Ignorance: Many people may simply lack information that would provide a meaningful context for coming to a well-grounded opinion.
2. Cost: An aggressive space program is expensive, and the money invested could be used by government for a better purpose, or returned to the taxpayer.
3. Irrelevance: However glorious space exploration might be for the engineers and astronauts who undertake it, there may be little or no significance for the daily lives of average citizens.
4. Discomfort: Like many other branches of science, astronomy and related fields contradict the traditional myths that have afforded comfort to individual people and that provided systems of values to organize successful societies.

These four can be combined, and it is entirely possible that all of them were at work, reinforcing each other in complex ways in the minds of various citizens. Opinion poll data can evaluate some of the four, but a very major social scientific research project would be required to disentangle their effects and determine their relative strengths. Here, the most we can do is provide analysis that documents and clarifies them to a moderate extent, adding more depth in later chapters.

Public *ignorance* of the subject matter of space science is profound and continues today, but was more extreme in the early days (Smith 2003). With respect to ignorance, I "conducted a tiny poll" on July 21, 1969, with just two respondents. To call it a "poll" is an exaggeration, but it was an informative experience. I watched the televised moment when Neil Armstrong took his famous step on the lunar surface, in the company of an uncle and an aunt. Both were college educated intellectuals, but neither was aware that the Moon lacked a breathable atmosphere.

Rather more substantial were two polls conducted in the United States roughly 6 months before Sputnik I and 6 months afterward, with 1,919 and 1,547 respondents (Swinehart and McLeod 1960). At the earlier point in time, the news had already carried stories about the plan to launch the first American satellite called Vanguard during a symbolic period of international focus on science, the International Geophysical Year 1957–1958, but only the launch of Sputnik made many in the general public take notice. In the earlier survey, only 21 % of respondents could suggest any purpose for such a satellite, compared with fully 64 % afterward. Of respondents to the later survey, 20 % cited competition with the Russians, and 17 % cited future possibilities. Scientific goals were mentioned by 27 %, but only 11 % could be specific about what these scientific goals were.

At that point in the history of social science, research about the impact of the mass media was at its height, especially concerning the debate about the role played by relatively well-educated opinion leaders that was introduced in the previous chapter in the discussion of the Martian invasion "panic" of 1938. One standard finding was that opinion leaders paid attention to somewhat different sources of information than did the average citizen. Today, many educated people get their news from Internet, and magazines are in serious decline as both a source of information and a business. But in 1957, according to one survey, 44 % of people who had attended college primarily got their news from magazines, compared with 39 % from newspapers, just 15 % from television, and 1 % from radio (Wade and Schramm 1969). Newspapers did not show much difference by education, 42 % of people who had not even attended high school primarily using this medium. But only 13 % of this least-educated group cited magazines, 34 % get their news from television, and 6 % from radio.

Given the dominance at that time of the NBC, CBS, and ABC networks, broadcast media may have promoted more narrow perspectives on spaceflight, than did magazines which then included not only *Life*, *Time* and *Newsweek*, but also a great variety of specialty periodicals. *Collier's* magazine famously promoted the spaceflight movement through many excellent articles in the first half of the decade. In that same 1957 survey, 38 % of magazine readers knew some scientific information about Earth satellites, compared with 22 % of newspaper readers, 16 % of TV viewers, and 10 % of radio listeners. After Sputnik, in 1958, these levels of awareness had all increased, to 47 % for magazines, 34 % for newspapers, 25 % for television, and 19 % for radio. Whether resulting from low education or low quality news media, ignorance varied and generally declined in the wake of Sputnik, but is a continuing issue even today.

Data from the January 1969 Harris poll #1877, available to anyone from the Odum Institute at the University of North Carolina, elucidate the *cost* issue

Table 2.1 First priorities to cut or keep among government programs in 1969

Government program	Cut (%)	Keep (%)
Aid to cities	5.1	5.4
Anti-poverty program	6.1	18.3
Space program	40.7	1.3
Subsidies for farmers	6.5	3.3
Aid to education	0.6	20.2
Medicaid	1.6	8.4
Anti-air and anti-water pollution programs	1.7	6.9
Welfare and relief	9.5	8.0
Building more highways	8.6	0.8
Financing the war in Vietnam	18.4	3.9
Anti-crime and law enforcement program	1.1	23.4

(www.irss.unc.edu/odum/home2.jsp). This was immediately after the flight around the Moon by Apollo 8, heavily covered in both print and broadcast mass media, and when the general public expected an actual lunar landing soon. Thus one would think that support would have been at a high point. With respect to cost, many of the opinion polls explicitly posed questions about space exploration in terms of the investment required, and one such question from polls administered after 1972 will be the focus of the following chapter. Most surveys asking about space program funding presented it as a somewhat separate question, one among many questions about funding for government programs, but answerable independently of other programs. This Harris poll asked a very different kind of question, focusing the mind of respondents on the trade-offs. The interviewer's questionnaire administration instructions stated it thus: "Now I want to give you this list (HAND RESPONDENT CARD 'B') of government programs. If *one* program had to be reduced, which *one* would you cut first?" Table 2.1 shows the results, plus the responses to a second question asking which program "you most like to see kept or even increased."

Of the 1,436 respondents willing to suggest which program to cut, fully 40.7 % put the space program on the chopping block, and only 1.3 % of the 1,444 willing to identify their top priority to save selected it. More than twice as many wanted to cut funding for the space program than the runner-up, which was the Vietnam war. This does not mean that they lacked all appreciation for it, but when presented with the harsh trade-offs for investment of public money, they became discouraged.

Another question in the same 1969 Harris poll illustrates the related issue of *irrelevance* to the every-day life of the respondent. It asked: "How would you rate the state of your health—would you say it is excellent, pretty good, only fair or poor?" Another question asked: "Do you favor or oppose the space project aim of landing a man on the moon?" Table 2.2 cross-tabulates these two, using the 1,252 respondents who had some kind of health insurance to reduce somewhat any tendency for the results to be dominated by cost considerations, and using this particular space question for the same reason, because it does not mention cost.

Table 2.2 Health and support for a landing on the Moon, 1969

Health	Favor Moon landing (%)	Not sure (%)	Oppose Moon landing (%)
Excellent	51.3	11.2	37.5
Pretty good	39.1	12.7	48.1
Only fair	26.1	14.6	49.2
Poor	12.0	20.0	68.0

The results could hardly be more striking, given the fact that many variables shape responses to survey questions. As health declines, support for a Moon landing drops from 51.3 % all the way down to 12.0 %. In later chapters we will examine the effect of age, which is complex but universally shows low support for space projects among the elderly, and they tend to face many health problems. People who are economically poor may also disproportionately have poor health, both as cause and effect of poverty. Whether because of low education or their appreciation for government anti-poverty programs, they may oppose space programs. But all those factors aside, I believe we can read something else into this table. People whose concerns are very much focused on their own personal condition, in this case ill health, will be far less interested in grand projects that are irrelevant to their own immediate needs.

Discomfort is a deep issue, and therefore difficult to disentangle from other issues. Consider the history of astronomy and physics. We revere the memories of Copernicus, Galileo, and Newton precisely because they taught us to think in new ways about the universe, ways that may not have been "natural" even though they illuminated the facts of nature. Humanity evolved in East African forests and savannahs, then spread around the globe sufficiently rapidly that our cognitive evolution could not keep pace with our geographic expansion. We see this in the difficulty humanity experienced in devising cultural structures to sustain societies much larger in scope than the tiny hunter-gatherer bands the biological evolution suited us for (Lévi-Strauss 1969). We also see it in how slowly humanity came to understand the shape and dynamics of the solar system.

While many myths have arisen about Stonehenge, having stood within its circle, and leaned against the Heel Stone, I can well believe that it was arranged to mark the apparent motions of the sun across the sky, fully 5,000 years ago. Some ancient Greek scientists appreciated the possibility that the Earth was round, but until navigators circled it in the much later Age of Discovery the general public conceptualized it the way their eyes testified: flat. Thomas Kuhn could title an influential book *The Copernican Revolution*, because humans instinctively assume their world is the center of the cosmos, and would feel uncomfortable to sense that their world is spinning around (Kuhn 1957). In a lecture I attended, innovative visionary Buckminster Fuller reported that he liked to stand looking at the horizon and fully sense how fast he was zooming through space, but most people might find this experience disconcerting or even terrifying. In so doing, they extrapolate from the narrow confines of their daily existence, where falling fast is extremely dangerous.

The most informative other example is the human reluctance to accept the full implications of Darwin's theory of evolution by natural selection from random genetic variations (Bohannon et al. 2010). In this case, unlike the situation with spaceflight, there exists an opposition movement, Creationism, which asserts God created us directly, for his own divine purposes (Numbers 2006). H. Porter Abbott has argued that reluctance to accept evolution is not merely a reflection of commitments to a set of ancient beliefs, but reflects a standard feature of human cognition. Abbott is a leading scholar of narrative, having written extensively about its roles and variations in literature. In one essay published in a collection about cognitive science, he argued that the theory of evolution from natural selection is unnatural because it is *unnarratable* (Abbott 2003). Human action is guided by purposeful thinking in terms of a narrative in which a person seeks a goal, must overcome a series of obstacles, progresses by assembling resources, recruiting allies, and seeking a path forward. Our brains evolved to think that way, because such thinking was advantageous in our mundane lives, but it is not conducive to any deep acceptance of the theory of evolution by natural selection from random variation.

The same may be true for the revolutionary perspectives that waited thousands of years until individuals with exceptional minds in rapidly changing societies could conceptualize them, namely Copernicus, Galileo, and Newton. Two kinds of poll data bear on the discomfort factor: (1) religious beliefs, and (2) stereotypes of scientists. They will be covered in Chap. 6, and the next task here is to consider the implications of the dawn of the Space Age, which began as a race.

2.2 A Growing Awareness

Long before the Second World War developed technologies on which real spaceflight could be built, small segments of the population were thinking about the possibilities. As we shall document in some detail later in this book, science fiction fans had been dreaming of voyages to other worlds since late in the nineteenth century. Some members of the public were interested in astronomy, for example reading *Sky and Telescope* magazine since it was first published in 1941, and indeed many of America's large telescopes had been funded by private donations. Further afield, professionals and students in all of the non-biological natural sciences would have learned something about the structure and dynamics of our solar system, and thus could see the plausibility of spaceflight. However, serious technical knowledge could also raise doubts, for example the awareness that the velocity a spaceship would need to achieve to escape the Earth's gravity, assuming reasonably enough that it launches from the surface, was fully seven miles a second.

The launch of Sputnik I from the Soviet Union, October 4, 1957, was a momentous public relations coup, and presented American president Dwight David Eisenhower with a multiplicity of problems. He was quite aware that both nations were far along on developing nuclear-armed intercontinental ballistic missiles, but had been constrained in how much he could reveal to the public. Anyone at all familiar with the technology would have known that the US could have launched the first satellite years earlier, had that been a priority. A good source of public information today was not available then: a page of the website of the Eisenhower Presidential Library devoted to Sputnik. Its introduction remarks:

> Rather than celebrating this momentous scientific feat, Americans reacted with a great deal of fear. The event came at a period near the end of the McCarthy communist "witch hunts," a time when schoolchildren were involved in "Duck and Cover" air raid drills, and citizens were encouraged to build their own civil defense shelters. It was widely believed that if the Soviets could launch a satellite into space, they probably could launch nuclear missiles capable of reaching U.S. shores.

That concern might not be moderated by the awareness that the US Air Force had long been in a position to destroy the Soviet Union. In December 1958, the US launched the first communications satellite, Project SCORE, broadcasting Eisenhower's own voice: "This is the President of the United States speaking. Through the marvels of scientific advance, my voice is coming to you from a satellite circling in outer space. My message is a simple one: Through this unique means I convey to you and to all mankind, America's wish for peace on Earth and goodwill toward men everywhere." All well and good, but what astonished me personally when SCORE achieved orbit was that it consisted of an entire Atlas intercontinental ballistic missile, minus the two extra engines it used at launch. Now, scanning the documents available on the Eisenhower Library's website, I especially note a statement about Sputnik from the National Science Board:

> The significance of the Soviet accomplishment in exploring outer space has been considered at length by the Board of the National Science Foundation. The Board regarded this as a great scientific and technical accomplishment; and urged that it be recognized as such. The Board further considered it an impressive demonstration of the strong position of Russian science and education... We must recognize that our nation's future rests in major degree upon the soundness of our system of education and our people's respect for scientific endeavor, based on an understanding of its importance in the modern world.

Such a statement has two kinds of meaning. First, its meaning is contained in the document itself, stating that Sputnik was a great scientific achievement, reflecting the strength of science and education in the Soviet Union. Second, it was an expression of the vested interest of the authors. The National Science Board is a committee of Presidential appointees, scientists themselves and mainly academics, with two responsibilities: (1) to carry out oversight of one US government science agency, the National Science Foundation, and (2) to advise the president on matters related to science. The Board's statement does not refer to rocket science narrowly, but to science more generally, and thus is advocacy for support of NSF, as well as contributing to the expansion of the National Advisory Committee for Aeronautics (NACA) into the National Aeronautics and Space Administration (NASA) over the following months.

Public reactions to Sputnik looked both forward, to what America might now accomplish in space, and backward. On October 10–15, 1957, just days after the launch, the Gallup Poll asked 1,573 American adults, "How long do you think it will be before men in rockets will reach the moon?" Respondents were about equally split, 52 % being willing to provide a definite prediction, and 48 % unwilling. On November 25, 1957, less than 2 months after Sputnik, Gallup included an open-ended question: "Where, specifically, would you put the blame, if anywhere, for letting the Russians get ahead of us in developing rockets and missiles?" The poll takers wrote down the respondents' verbal answers, and Gallup collected phrases into groups during a rough analysis of these non-quantitative data:

The president, Eisenhower, the White House
Leaders of Government, General; no specific individual group, or party mentioned; the politicians
Congress, Senators, Congressmen (no reference to appropriation of money)
The Republican Administration, the administration, the present administration, the government
(Congressional) cut-back in defense budget, not enough money, curtailment of funds by (Congress)
Charles Wilson, Defense Department
Inter-service rivalry, jealousy between forces, competition between Army, Navy, Air Force
Preceding Administrations, Roosevelt, Truman administration
All Americans, everyone is to blame (no mention of why or how)
Our complacency, smugness, cocksureness, neglectfulness, etc., not enough attention paid to business, too lazy
Lack of major party cooperation, Dems and Reps fighting each other
Russian espionage, spies got information, our know-how "leaked out," poor security
Not enough scientists, good scientists
Our failure to get the most (best) German scientists
Inadequate educational preparation for science, not enough stress on education
U.S. gave away too much information, too trusting
Poor planning, mismanagement, not enough emphasis on rockets, missiles
Restriction on scientists
No one is to blame
Doubts, does not believe Russia is ahead.

The question emphasized the word *blame*, and respondents may not have felt inclined to argue against an expensive American space program. People were not, for example, saying, "The Soviet Union was able to launch Sputnik because it has a totalitarian form of government, and therefore can waste money on worthless projects that a free democracy should ignore." Of course, the serious shock among American leaders about Sputnik was not so much about the Soviet decision to launch an Earth satellite, but their technical ability to do so.

The wider world was an attentive audience for the Space Race, from the launch of the first Sputnik until the last Apollo landing. Of course, citizens of the Soviet Union took pride in the early achievements. Some may have considered them proof of the superiority of Marxism, and others would have experienced them in more nationalistic terms. More widely, they could draw optimism about their own

Table 2.3 Percentage differences thinking Russia was ahead of the United States after Sputnik

Nation	Scientific discovery		Military strength	
	November 1957 (%)	October 1958 (%)	November 1957 (%)	October 1958 (%)
Great Britain	38	−13	31	15
West Germany	−4	−21	−15	−1
France	38	14	8	9
Italy	14	−3	−12	−15

futures, just a dozen years after the conclusion of the Second War that had been fought on their own territory and cost the lives of perhaps twenty million of their fellow citizens. In 1994, 2,400 residents of Russia gave their reaction to a list of prominent events from the previous half century, one of them being the launch into orbit of a dog named Laika in Sputnik II, November 3, 1957. This proved to be one of the best-known events, especially so among respondents who were children at the time Laika became the first space traveler (Schuman and Corning 2000).

People in many nations were impressed by the first Soviet satellites. Polls administered in four nations allied to the United States asked "All things considered, do you think the U.S. or Russia is ahead in scientific development at the present time." A second question asked the same for "total military strength." For example, in November 1957, only 20 % of British respondents thought the US was ahead, while fully 58 % thought Russia was ahead, a difference of 38 % points in favor of Russia. Table 2.3 shows the percentage differences across the four US allies at two dates, the second being a year after Sputnik when the US had also launched satellites. Minus signs indicate the view that the US was ahead of the USSR (Almond 1960).

It is hard to know when pollsters are asking leading questions that implicitly urge to respondents to answer in a particular way. Yet perhaps many people really did think that scientific discovery was a unitary phenomenon that could be illustrated by just one example, let alone the rather distorted notion that being able to launch satellites on converted military missiles was "scientific discovery." The first significant discovery, the van Allen radiation belts around the Earth, was in fact made by the first American satellites, not the first Soviet ones. A reasonable argument could be made that the ability to launch satellites required excellence in several fields of engineering, which could not be achieved without excellence in several areas of science.

A report prepared for the US Congress that reflected upon Sputnik in 2009 noted, "The United States faces a far different world today than 50 years ago. No Sputnik moment, Cold War, or space race exists to help policymakers clarify the goals of the nation's civilian space program." (Stine 2009) Yet it could be said that organization leaders saw Sputnik less as a clarification than as an opportunity, because it provided ammunition for some of them to promote goals they were already committed to.

2.3 Exploitation of the Situation

As the previous chapter noted, public opinion develops though complex processes of communication and influence, which leaves it vulnerable to manipulation by social movements and interest groups. In the context of the Sputnik Panic, three such entities can be readily identified: the Democratic Party, the military-industrial complex, and various reform movements within the educational establishment. Without taking a political stand, or expressing sympathy or antagonism toward any of these three, their roles in shaping the meaning of spaceflight can be outlined.

The November 25, 1957, Gallup Poll documented a fairly widespread sense that President Eisenhower and his Republican administration had failed to anticipate and then preempt the impact of the early Russian satellite launches. Quite reasonably, Sputnik raised concerns that the USSR might be building a vast fleet of intercontinental ballistic missiles to attack the United States, although from the Soviet perspective such a missile armada might be only a counterbalance to the huge American superiority in bomber aircraft, leveraged by the fact that the US had many airbases within striking distance of Mother Russia. The Democratic Party began to exploit this situation, asserting that a missile gap existed because the US was obviously far behind the USSR.

While the missile gap theory was reasonable, it also served political purposes. Given how close the vote counts in the 1960 presidential election turned out to be, it is conceivable that the missile gap hysteria gave John F. Kennedy the victory. The loser, Richard Nixon, gained the office 8 years later, and it boggles the mind to contemplate how history more generally might have been different, had he become president at his first attempt, and before the US had gotten embroiled in the Vietnam War. There is good reason to believe the Apollo Program would not have been launched had the Democrats not asserted the missile gap and Nixon had become president in 1960.

In retrospect, it is clear that the missile gap did not exist (Dick 1972). The United States had considerable flexibility in which aspects of its vast military establishment would play what role in deterring Soviet aggression, chiefly in Europe where it was most feared. But the early Soviet ICBMs were too costly, unreliable, and few to outweigh the deterrence already possessed by the US, who had the technical and economic ability to field better missiles faster. The rush to deploy the first ICBMs promoted the kind of rocket technology useful for spaceflight, namely liquid-fuel rockets, but already 5 years after Sputnik the solid-fuel Minuteman ICBM was going into service, more efficient and easier to hold in readiness. Among the factors making Minuteman and the submarine-launched Polaris missiles feasible were advances in nuclear weapons and guidance technologies that produced much smaller warheads. The Mercury and Gemini astronauts were launched on modified Atlas and Titan ICBMs. Thus, one aspect of the Space Race was technical, to promote spaceflight before some of the military technologies had been perfected.

It would be an oversimplification to say that the Moon program was funded because it served the interests of the "military-industrial complex," yet there is some truth to this claim. The very concept sounds like bombastic nonsense uttered by a sociologist, or even nasty Marxist propaganda. But it was introduced into American intellectual life by one of the very most trusted political and military leaders that America has ever produced, president and general Dwight David Eisenhower. He did so in his extremely prominent farewell address to the nation, January 17, 1961.

Interested readers may consult the several versions of Eisenhower's intellectually rather deep lecture, which I watched with amazement as he originally gave it on television, on the website of the Eisenhower Presidential Library. I was especially astonished when he said, "In the councils of government, we must guard against the acquisition of unwarranted influence, whether sought or unsought, by the military-industrial complex. The potential for the disastrous rise of misplaced power exists and will persist." He specifically warned that science policy could be horribly distorted by the military-industrial complex, but also mentioned the possibility that science itself might gain too much influence. At the time, I was aware that he had not vigorously supported an aggressive manned space program, and that his vice president, Richard Nixon, had lost the election that decided Kennedy would be his successor in part because of claims that the Republicans had allowed that "missile gap" to exist between the USSR and USA. So I understood Eisenhower to be defending his past indifference toward science, and only in later years did I come to appreciate how well he had framed very legitimate issues.

It is very clear that the American educational establishment jumped on Sputnik, as a rhetorical resource to advance its own agenda, not infrequently sacrificing intellectual accuracy for political expediency. For example, in the April 1958 issue of *The Journal of Higher Education*, Thomas N. Bonner wrote, "Mr. Teller, certainly one of the most knowledgeable and moderate of the scientists who have commented, testified emphatically that it will take us a minimum of 10 years of hard work a full speed before we can hope to draw abreast of the Soviets in basic scientific research and knowledge affecting missile and rocket development (Bonner 1958)." The Teller of whom Bonner speaks was Edward Teller, "father of the hydrogen bomb" who later promoted the space-based Strategic Defense Initiative, which we shall cover in Chap. 5, and who is generally recalled as a "hawk" and not a moderate. Bonner's fundamental argument was not really about Sputnik, although that was the superficial focus of his essay, but about American anti-intellectualism, which many university scholars decried during the years around 1960 (Hofstadter 1963).

Bonner referred to an educational crisis in America, and other authors used the same journal to publish essays demanding reform of curriculum and student support (Hilberry 1958; Graham 1959). Such a crisis may have existed, but Sputnik was not evidence of it, and contrary to Teller's claim the Soviet Union was not ahead of the United States in science. On October 4, 1997, exactly 40 years after the launch of Sputnik I, the Center for Science, Mathematics, and Engineering Education held a symposium at the National Academy of Sciences reflecting upon that great event. Contributors noted that reform movements in science and mathematics education already were active in 1957, but Sputnik gave them a big boost for perhaps two decades (www.nas.edu/sputnik/index.htm). The excessively abstract New Math movement, which did not take much account of the psychology of child development, came under criticism, so not all the reforms facilitated by Sputnik may have been good (Raimi 2012). One good feature of the Sputnik response was a closer connection between educators and scientists, a bond that may have weakened again subsequently. We have already seen that better-educated people tend to be more supportive of spaceflight, but in the following chapters we will see much evidence that the link between the space program and science education is a tenuous one.

If I may share a personal observation, relevant to education during the Space Race, I was caught up in it immediately after Sputnik. The Choate high school I attended got a phone call from the producers of the tremendously popular television quiz program, *The $64,000 Question*, asking for students who would become contestants to prove how excellent the American educational institution really was. I and a younger boy were sent to New York, and I am not sure how well I did on a written test about the space program, although I know I got one orbital mechanics formula wrong. But I certainly flunked the interview, in which I forcefully stated my belief that the American public educational system was indeed dreadful, and was not invited to become a contestant. Only subsequently, when this quiz show was revealed like some others to be manipulating outcomes in favor of popular contestants, did I realize my tactical mistake.

Perhaps popular opinion, and public understanding of science, belong to the context in which technological innovation takes place, rather than being central to the process of innovation itself. In my book *The Spaceflight Revolution*, I outlined a theory that was based on the historical work of previous scholars on the Space Race, but sought to abstract a model that applied to other examples as well. The model concerned social interaction among three actors in a social system, which could be individual leaders or organizations:

1. The *spaceman*: A leader in the spaceflight social movement who seek resources to advance its goals.
2. The *patron*: A leader in the larger society who has considerable resources at his disposal and is relatively free to spend them without external control.
3. The *opponent*: A leader who is locked in fierce competition against the patron.

For example, the spaceman might have been German rocket engineer Wernher von Braun, the patron might have been Adolf Hitler, and the opponent was the enemies of Germany in the Second World War. In the model, the opponent gains an advantage over the patron, and the patron who is aware of this seeks a countermeasure without quickly finding one. The spaceman can then go to the patron and sell his favorite spaceflight advance as a solution, rightly or wrongly, to the patron's problem of momentary inferiority to his opponent. Here is how the model applies to the case of the Apollo Project, specifically the Saturn I launcher which was the predecessor to the Saturn V that sent men to the Moon:

CASE 14:

PURPOSE: Development of superbooster suitable for lunar and space station missions
SPACEMAN: Von Braun and others in NASA
OPPONENT: The Soviet Union, the Republican Party

PATRON'S INFERIORITY: American spacefaring inferiority as evidence by Russian Sputniks and Gargarin's orbital flight; Kennedy's "New Frontier" in trouble (partly because of the Bay of Pigs fiasco) and needs new visionary boost to regain credibility (Bainbridge 1976).

This was Case 14 of 15 I outlined, the Space Shuttle being the last one, and von Braun's personal pitch to Hitler about increasing investments in the V-2 rocket program was Case 7. Again, the fundamental analysis of my Case 14 had been done by other scholars, especially Vernon van Dyke and John Logsdon (van Dyke 1964; Logsdon 1970). Perhaps this theory gives too much credit to the spaceflight social movement and too little to the military-industrial complex. But, in the context of this book an important point is that neither the general public nor wider scientific community played a decisive role. The general public is the audience for the posturing of politicians, who use scientists and engineers as supporting players in their drama. This does not negate the significance of public opinion, but locates it where it belongs, as an arbiter of the meaning of spaceflight, rather than as its instigator.

2.4 The Vietnam War

During the period of the Apollo Program, the United States fought a politically divisive war in Vietnam, beginning during the Kennedy and Johnson administrations that represented the Democratic Party, and then concluding painfully during the Republican administration of Richard Nixon. Harris poll #1718 was administered in April 1967, and provides a good glimpse of the political implications near the middle of the Apollo Program, about a year and a half before the first manned mission, Apollo 7, which tested the vehicle in Earth orbit. This was also several months before the Tet offensive in the war that galvanized popular opposition.

The questionnaire included three fixed-choice questions about spaceflight, which were also included in several other Harris polls of that general period. The first put the program in the context of government funding: "It could cost the United States $4 billion a year for the next 10 years to finally put a man on the moon and to explore other planets and outer space. All in all, do you feel the space program is worth spending that amount of money on or do you feel it isn't worth it?" Of the 1,177 respondents who answered this question in April, 32.9 % felt the investment would be worth while, 54.5 % felt it would not be worth it, and 12.6 % were unsure. Thus, a majority of the public was against an ambitious space program.

The second question put the issue in the content of international competition: "If the Russians were not in space, and we were the only ones exploring space, would you favor or oppose continuing our space program at the present rate?" Now, only 29.0 % were in favor, 60.4 % opposed, and 10.6 % unsure. This indicates that competition with the Soviet Union was indeed a factor in shaping public opinion about spaceflight, although not really decisive because it had not tipped the balance over to a majority supporting a vigorous space program.

The third question was focused more narrowly on the Apollo Program: "Do you favor or oppose the space project aim of landing a man on the moon?" Now, 41.8 % were in favor, 47.0 % were opposed, and 11.2 % not sure. This item differed from the others in two ways: (1) It did not emphasize the great cost. (2) It focused on the single goal of reaching the Moon. The fact that support was somewhat stronger for this item suggests that the public was not prepared to give NASA a blank check without a clear understanding of what would be accomplished, and that an expedition to the Moon had somewhat clear and positive meaning. However, none of the three questions indicated a majority in support.

Given that Apollo was a Kennedy initiative, continued by Johnson, it may have been more popular among members of their own party, the Democrats. One could argue that aspects of Republican ideology might have favored it as well. The situation is complicated by the ambiguous political meaning of the Vietnam War. Now, half a century later, it is hard to guess how things might have turned out had the 1960 or 1964 elections gone the other way, but thankfully the American stumbling around in Vietnam did not lead to nuclear war. By the 1968 election, substantial opposition to the war had arisen in the Democratic party, and it was left to the subsequent Republic administration to deal with the problem in its own incompetent manner. The objective point to draw from this superficial overview is that both Apollo and Vietnam were connected to the Democratic Party in 1967, so political affiliation may have shaped public opinion about both in a similar fashion.

The first thing to note about political affiliation in the Harris poll #1718 data is that Democrats greatly outnumbered Republicans. Of the 1,164 respondents who answered the first spaceflight question and gave their political orientation, only 30.9 % were Republicans, fully 52.3 % were Democrats, and 13.4 % called themselves Independents. Of these groups, the one showing most support for space funding were the Independents, with 40.4 % calling a $4 billion annual investment "worth it" and 47.4 % "not worth it." Second place in terms of space support was indeed held by Democrats, 33.8 % being in favor versus 54.4 % against. Among Republicans, 28.6 % felt the program was worthwhile, compared with 58.9 % who did not.

In the following chapter we will consider political orientation again, and it is a complex variable. Some Republican opposition may stem from opposition to big spending by government and some from association of Apollo with the Kennedy-Johnson administration. Give that spaceflight is literally "out of this world," its connection to mundane political issues may be unstable, changing markedly over time as the public assigns different meanings to it. Removing competition with the Russians from the equation, as the second Harris question does, reduces spaceflight support in all three political categories, to 32.7 % among Independents, 30.9 % among Democrats, and 24.7 % among Republicans.

All three political categories respond more favorably when the clear goal is reaching the Moon, without an indefinite financial commitment. Among Independents, now a plurality of 48.1 % favored the program, compared with 45.5 % who opposed it. The difference went the other way for Democrats, 42.2 % favoring and 45.5 % opposing. And the difference was even greater among Republicans, 39.3 % in favor and 50.4 % opposed. This suggests that many

Table 2.4 Alternative strategies for the Vietnam war

Opinion	Cases	Worth $4 billion per year (%)	Favor even without Russia (%)	Favor Moon program (%)
Point of view about war in Vietnam				
I disagree with present policy. We are not going far enough. We should go further, such as carrying the ground war into North Vietnam	178	30.9	27.8	40.4
I agree with what we are doing but we should increase our military effort to win a clear military victory in South Vietnam	463	40.0	33.8	50.1
I agree with what we are doing, but we should do more to bring about negotiations such as a cease fire	308	33.1	29.5	42.5
I disagree with present policy. We shouldn't be there. We should pull our troops out now	136	17.6	16.1	22.8
Do you favor or oppose the use of atomic ground weapons in the fighting in Vietnam?				
Favor	113	36.0	31.0	47.6
Oppose	217	33.3	29.2	41.9

Republicans may indeed have associated Apollo with the Dcmocrats, but even the Democrats themselves were not strong supporters. With this background, we can now see what may be revealed by comparing three spaceflight questions with two that Harris asked about the war.

One especially interesting item asked respondents to select one of the four policies listed in Table 2.4. The numbers of cases given in the table are the number giving this response and answering the first spaceflight item, and the numbers are similar for the other two spaceflight items. The patterns are the same across all three space items. Those content with current policy in the war are most content with the Moon program. Those who are somewhat discontented and want moderate changes in policy, either more militaristic or more willing to negotiate, are less enthusiastic, about equally so. Those who were strongly against the war and wanted the US to withdraw from Vietnam were very unenthusiastic about the Moon program. Given that many of those "peaceniks," as they were derisively

Table 2.5 Preparations for nuclear war

Opinion	Ought to be done (%)	Not sure (%)	Should not do (%)	Space item correlation (tau-b)
Build up a system of anti-missile defenses	60.1	16.7	23.2	0.29
Increase the number of airplanes which carry nuclear warheads	48.1	19.4	32.6	0.39
Increase the number of nuclear warhead long-range missiles	51.5	20.9	27.5	0.38
Increase the number of men in the U.S. armed forces	31.1	17.0	51.9	0.33
Give NATO a real capability of waging nuclear warfare	25.1	27.3	47.6	0.34
Convert the space program into a system of nuclear weapon space stations	25.1	26.4	48.6	–

called by those who associated anti-war sentiments with Communism and thus indirectly with Sputnik, would have leaned toward the Democratic Party, they may have reduced the party differences in support for the space program, below what they would have been without the Vietnam War.

At the bottom of the table we see an item that was asked of only a subset of respondents, concerning the use of nuclear weapons in the Vietnam War. They had not been used in the Korean War, for example at the point when China sent its forces into help the North Koreans after their armies had retreated in the face of General McArthur's classic encirclement maneuver at the Battle of Inchon. The question about use of nuclear weapons does not show big differences, but does remind us of the remarkable range of views about them that existed decades ago.

Table 2.5 is based on Harris poll #1900, administered to 1,544 respondents in December 1968, which included some rather chilling questions about "areas where it has been suggested the U.S. military defenses be strengthened," including one about full militarization of the space program. This was a period in history when widespread nuclear war appeared to be quite possible, and strategists debated issues like striking the enemy first, basing peace on mutually assured destruction if war came, and other ideas that seem at best hazardous if not downright immoral to contemplate (Kahn 1960, 1962).

The first item, "anti-missile defenses," came back into prominence in the 1980s with the Strategic Defense Initiative, and 60.1 % of respondents were in favor, versus 23.2 % who were against. It has a correlation (tau-b for two non-parametric items with the same number of response categories) of 0.29 with the last idea in the list, of militarizing the space program. The other items have higher correlations with the space militarization item, undoubtedly reflecting the fact that relatively pacifist respondents could well support a system of anti-missile defenses while being opposed to more aggressive space-related technologies. The Vietnam War took place long ago, much closer to the Second World War than to today, but the range of public opinions about the space program that consolidated during that period persists.

2.5 The Unlucky Apollo

In April 1970, an oxygen tank on the service module of Apollo 13 exploded, as the craft was on its way to the Moon, initiating a remarkably complex but ultimately successful effort to return the crew safely. A week later, Harris Poll #2025 asked 1,520 American adults how worried they were "whether the men in the spaceship would get back to earth," and 54.5 % admitted to being very worried. Another question asked, "Do you expect that on one of the space shots an accident will take place and the astronauts won't get back alive, or do you think that probably won't happen?" Fully 71.2 % anticipated a future fatal accident. Frankly, it is somewhat difficult to frame exactly what these two questions were really measuring. The second question did not specify when a future accident might happen, and given enough years of exploration a fatal accident was bound to occur. In fact, in the context of NASA's manned spaceflight program, it took 16 years until the Challenger disaster for the question to receive an objective answer.

The poll also asked variants of Harris's standard "worth it" items, beginning with: "Getting to the moon cost 4 billion dollars a year for 9 years. Do you feel landing a man on the moon was worth spending that amount of money, or wasn't it worth it?" Just 38.3 % of respondents felt the moon program was worth it, compared with 56.3 % who said it had not been. The other item was the same one from the April 1967 poll about whether space exploration was worth 4 billion dollars a year for 10 years. It showed a slight drop in support, just 29.5 % feeling it was worth it, compared with 32.9 % 3 years earlier.

What makes the 1970 poll really interesting was that it also included many questions about ordinary aviation, stimulated by the fact that the gigantic Boeing 747 had gone into regular service just 3 months earlier. This allows a glimpse into the extent to which people conceptualize spaceflight as an extension of the aviation they experience themselves. One question asked if the respondent had flown with the past year, and another given only to those who had not flown that recently asked whether they had flown within 5 years. Combining the data from these two items identifies 615 respondents who had flown within 5 years, and 817 who had not. Now this at least partly reflects social class, and we would need a much richer dataset to determine conclusively why the experience of flying should correlate with spaceflight attitudes. But for present purposes we can conjecture that respondents who had flown would conceptualize spaceflight as an extension of an experience they themselves had, thus being more comfortable with the idea.

Of those who had flown, 55.6 % were worried during the Apollo 13 mission, compared with 54.4 % who had not flown, essentially no difference. On the likelihood of a future accident, the percentages were slightly more different, 69.6 and 73.1 %, respectively. But a really big difference appears for the two questions about whether space exploration was worth the money invested. Of those who had flown in the past 5 years, 50.7 % felt getting a man to the moon had been worth it, compared with only 30.5 % of those who had not flown. The percentages for the more general question about investing in space gave lower but also different percentages, 38.2 and 23.8. A battery of 16 questions focused on the Boeing 747, eight of which are tallied in Table 2.6.

Table 2.6 Attitudes about the Boeing 747 and the space program

Agree-disagree statement	Percent agree (%)	Of those who agree with the statement			
		Moon worth it (%)	Program worth it (%)	Very worried (%)	Accident (%)
Bigger and better planes such as the 747 are a sign of sound technological progress	64.4	45.2	35.5	58.1	71.7
At a time when the airlines are faced with rising prices, the 747 will save the airline industry	36.2	44.9	37.2	59.1	73.7
Everyone in the 747 has a clear view of a much bigger screen for the first run movies	47.3	42.7	33.8	58.1	72.5
In the 747, the seats are larger and the passengers can fly with much more comfort	70.2	41.5	32.6	56.2	71.4

(continued)

Table 2.6 (continued)

Agree-disagree statement	Percent agree (%)	Of those who agree with the statement			
		Moon worth it (%)	Program worth it (%)	Very worried (%)	Accident (%)
The 747 has been rushed into service before all of its mechanical equipment was properly tested	16.6	38.8	32.0	58.8	77.2
The new plane's engines are so powerful that they can create a serious noise problem around the airports where they land and take off	46.9	35.6	27.4	58.5	74.1
The 747 gives off too much smoke which will cause it to pollute the air in cities where it lands and takes off	36.5	34.4	25.4	59.9	75.0
The old jets served the needs of flyers and new bigger planes just are unnecessary	18.8	30.9	21.6	56.9	77.6

The eight statements about the 747 are arranged in terms of descending percentages who felt sending men to the Moon had been worth it. The first item describes the 747 as "sound technological progress," and the last one calls it "unnecessary." Among people who agreed with the first of these, 45.2 % felt the Moon landings were worth it, compared with only 30.9 %. The two extreme 747 items are indirectly measuring whether people felt the Apollo Program itself was "sound technological progress" or "unnecessary."

2.6 A Thoughtful Retrospective

As historian Roger Launius has repeatedly reminded us, at no point during the period that began with the first Sputnik in 1957 and ended either with the last Moon landing in 1972 or the joint American–Russian Apollo-Soyuz test project in 1975, did a majority of the American public support increased funding for space exploration. This was well known at the time, because the public opinion polls themselves received considerable publicity, and scholars commented about it soon afterward. But Launius has contributed more than just a reminder; he has offered analysis of the meanings of spaceflight for the American public at different points in time, especially during the Space Race and Apollo years. A brief consideration of the analysis in four of his journal articles is warranted here, not merely to give him credit for these contributions but to consider the range of hypotheses he offers. When Launius argues in favor of one of these ideas, I believe, he is primarily clarifying them through examples, but also offers confirming evidence in many instances.

Most obviously relevant is "Perceptions of Apollo: Myth, Nostalgia, Memory or All of the Above?" (Launius 2005). In this 2005 retrospective, informed by his knowledge of public opinion research but not delving deeply into it, Launius offers six plausible meanings that the Apollo program may have had, here described in my own terms:

1. A Mythic Event
   Much cultural anthropology supports the idea that every functioning culture possesses myths that define the society's values, express a fundamental conception of reality through poetic language, and assert that unified social life has transcendental meaning. For Americans, the Apollo program may have expressed a national ethic of heroism, nobility, and innovation.
2. The Agrarian Myth of the Frontier
   Just as Americans pioneered westward in the nineteenth century, they imagined they were pioneering upward in the twentieth. The concluding chapter of this book will explore the "space frontier" concept in some depth, but it must be noted here that the analogy between the Apollo program and the frontier of the Wild West is actually a poor one, if very popular, because the space program is a centrally controlled government enterprise, whereas the traditional frontier was individualistic and liberating.
3. Skepticism Unbound
   Some fraction of the American population has long doubted the promises of government leaders and of technological advance, so a pessimistic appraisal always existed in the public mind, as well as whatever positive enthusiasm the spaceflight social movement could generate. Launius presents this perspective as a reaction to disappointment that the agrarian myth of the frontier was not in fact fulfilled by Apollo, but in some quarters disillusionment was not necessary because the original illusions promoting Apollo were not universally accepted.
4. A Bastion of Concentrated Power
   Apollo was the child of the military-industrial complex, born in the technocratic Kennedy administration that believed every problem had a technical

solution, if managed wisely by the federal government. Given that all the early human space missions launched using adapted intercontinental ballistic missiles, and the later ones modified the military technology only moderately, it is easy to conceptualize Apollo as the upper stage of a governmental bureaucracy, the lower stage of which was war machinery.

5. A Grand Vision for the Future
   Incorporating an exploration mythos rooted in the European Age of Discovery more than in the American frontier myth, Apollo was a symbol of progress more generally. It was timely, because the dismal period of the Great Depression and Second World War had been followed by two decades of national security and economic development, but it was unclear how progress could continue after the earlier traumas had been completely healed.
6. Nostalgia
   Now that more than four decades have passed since the last human voyage to the Moon, many people romanticize Apollo and seek to recapture its lost idealism. Perhaps many people forget the chaos of the 1960s, or wish we could find in our past some shining light to illuminate our path forward, but nostalgia implies a lack of realism, as well as a degree of sadness.

In 2003 Launius had published "Public Opinion Polls and Perception of US Human Spaceflight." After presenting some of the poll data, he observed:

> These statistics do not demonstrate an unqualified support for NASA's effort to reach the Moon in the 1960s. They suggest, instead, that the political crisis that brought public support to the initial lunar landing decision was fleeting and within a short period the coalition that announced it had to reconsider their decision. It also suggests that the public was never enthusiastic about human lunar exploration, and especially about the costs associated with it. What enthusiasm it may have enjoyed waned over time, until by the end of the *Apollo* program in December 1972 one has the image of the program as something akin to a limping marathoner straining with every muscle to reach the finish line before collapsing (Launius 2003a).

In another 2003 journal article, Launius noted that the general public has always overestimated the fraction of the federal budget invested in space exploration, so conceivably informing the public better might increase support (Launius 2003b). In 2012 he suggested three lessons from the Apollo years that might help guide NASA in setting a new course after the space shuttle: (1) There was less popularity than Apollo's promoters expected. (2) The technological challenges were greater than initially perceived. (3) Promising too much can incur political costs (Launius 2012).

2.7 To Win Is to Lose

The Space Race illustrates the contingent quality of history, in which developments of great significance may depend upon the chance conjunction of multiple independent trends. We have already mentioned the 1960 election and the Vietnam War as cases in point, unrelated to spaceflight but affecting its course, but suppose the

Apollo Program had failed spectacularly on technical grounds. Three astronauts did die in Apollo 1, but in a test on the ground rather than in space, and three astronauts nearly lost their lives in Apollo 13. Suppose for example that the Apollo 1 fire had not ignited, and the safety improvements to the capsule had not been made. Then Apollo 11 might have been the one destroyed, perhaps during one of its lunar maneuvers, and imagine also that the Apollo 13 accident happened on Apollo 12, but with a fatal outcome. President Nixon could quite easily have called Apollo another Kennedy-Johnson folly and cancelled it. This point of this string of counterfactual assumptions is to set the stage for another possibility. In the late 1970s the Soviet Union could have reached the Moon first, causing the US to race it to Mars.

Remarkable events have more power when the lack of powerful public opinion commitments leaves open the possibility that politicians or social movement leaders will determine the course of history, using these events as rhetorical tools. This observation takes on more meaning in the context of a sociological theory that is exceedingly uncomfortable with this insight, technological determinism, which was introduced in the previous chapter. William F. Ogburn assumed that public opinion was significant only in the difficult process of adaptation to technological changes, not in creating them. If he was right, then the weak support given spaceflight by the general public could represent cultural lag, a reactionary pathology that will be cured by the passage of time.

So today the Space Race may have few lessons for the future of space exploration, unless we reach a second point in history at which rapid technological advance is possible, and unusual political conditions create an opportunity to invest sufficient funds in achieving it (Schulman 1975). The space race ended four decades ago, and we have not seen favorable conditions like those during that time. However, it is possible that American culture has been gradually eliminating the cultural lag that inhibits an alternate approach based on public enthusiasm for sustained if gradual progress. The best way to explore this possibility, while also identifying multiple meanings spaceflight may have acquired, is to examine public opinion data from a single, consistent study that covered the full set of years between the end of the Space Race and the analysis of the data here.

References

Abbott, H. P. (2003). Unnarratable knowledge: The difficulty of understanding evolution by natural selection. In D. Herman (Ed.), *Narrative theory and the cognitive sciences* (pp. 143–162). Stanford, California: Center for the Study of Language and Information.

Almond, G. A. (1960). Public opinion and the development of space technology. *Public Opinion Quarterly, 24*(4), 553–572.

Bainbridge, W. S. (1976). *The spaceflight revolution* (p. 80). New York: Wiley-Interscience.

Bohannon, J., Gregory, T. R., Eldredge, N., & Bainbridge, W. S. (2010). Spore: Assessment of the science in an evolution-oriented game. In W. S. Bainbridge (Ed.), *Online worlds: Convergence of the real and the virtual* (pp. 71–86). London: Springer.

Bonner, T. N. (1958). Sputniks and the educational crisis in America. *The Journal of Higher Education, 29*(4), 178.

Commissioned papers from the National Academy of Sciences symposium Reflecting on Sputnik: Linking the Past, Present, and Future of Education Reform. Retrieved January 3, 2014, from www.nas.edu/sputnik/index.htm.

Dick, J. C. (1972). The strategic arms race, 1957–1961: Who opened a missile gap? *The Journal of Politics, 34*(4), 1062–1110.

Goodwin, H. L. (1965). *The images of space*. New York: Holt, Rinehart and Winston.

Graham, V. E. (1959). Sputnik and student aid. *The Journal of Higher Education, 30*(6), 331–334.

Gusfield, J. R. (1963). *Symbolic crusade: Status politics and the American temperance movement*. Urbana, Illinois: University of Illinois Press.

Hilberry, C. B. (1958). Sputnik and the universities. *The Journal of Higher Education, 29*(7), 373–380.

Hofstadter, R. (1963). *Anti-intellectualism in American life*. New York: Knopf.

Kahn, H. (1960). *On thermonuclear war*. Princeton, New Jersey: Princeton University Press.

Kahn, H. (1962). *Thinking about the unthinkable*. New York: Horizon.

Kuhn, T. S. (1957). *The Copernican revolution: Planetary astronomy in the development of Western thought*. Cambridge, Massachusetts: Harvard University Press.

Launius, R. (2003a). Public opinion polls and perceptions of US human spaceflight. *Space Policy, 19*, 168.

Launius, R. (2003b). Evolving public perceptions of spaceflight in American culture. *Acta Astronautica, 53*, 823–831.

Launius, R. (2005). Perceptions of Apollo: Myth, nostalgia, memory or all of the above? *Space Policy, 21*, 129–139.

Launius, R. (2012). Planning the post-Apollo space program: Are there lessons for the present? *Space Policy, 28*, 38–44.

Lévi-Strauss, C. (1969). *The elementary structures of kinship*. Boston: Beacon Press.

Logsdon, J. M. (1970). *The decision to go to the Moon*. Cambridge, Massachusetts: MIT Press.

Numbers, R. L. (2006). *The Creationists: From scientific creationism to intelligent design*. Cambridge, Massachusetts: Harvard University Press.

Raimi, R. A. (2012). The new math. In W. S. Bainbridge (Ed.), *Leadership in science and technology* (pp. 861–868). Thousand Oaks, California: Sage.

Schulman, P. R. (1975). Nonincremental policy making: Notes toward an alternative paradigm. *The American Political Science Review, 69*(4), 1354–1370.

Schuman, H., & Corning, A. D. (2000). Collective knowledge of public events: The Soviet era from the great purge to glasnost. *American Journal of Sociology, 105*(4), 913–956.

Smith, H. A. (2003). Public attitudes toward space science. *Space Science Reviews, 105*, 493–505.

Stine, D. D. (2009). *U.S. civilian space policy priorities: Reflections 50 years after Sputnik. Congressional Research Service*, February 2, (p. 5). Retrieved March 2, 2014, from opencrs.com/document/RL34263/.

Swinehart, J. W., & McLeod, J. M. (1960). News about science: Channels, audiences, and effects. *The Public Opinion Quarterly, 24*(4), 583–589.

van Dyke, V. (1964). *Pride and power: The rationale of the space program*. Urbana: University of Illinois Press.

Wade, S., & Schramm, W. (1969). The mass media as sources of public affairs, science, and health knowledge. *The Public Opinion Quarterly, 33*(2), 197–209.

Odum states: "Research publications, news or magazine articles and radio or television broadcasts employing statistical summaries of the Odum Institute data should give an appropriate citation to the Institute as the source of the data." All the Harris data analyzed for this book came from the Odum Institute, and I deeply appreciate the contribution to social science and public awareness provided by Harris and Odum.

Chapter 3
The General Social Survey

In 2012, 22.6 % of adults in the United States wanted funding for the space exploration program increased; 44.4 % wanted it kept the same, and 33.0 % wanted it reduced. For every 100 persons who wanted funding increased, 146 wanted it decreased, so if the American government were managed through a system of direct democracy based on public referendums about all important topics, on balance the nation's investments in space exploration would decline. Apparently, many Americans value other things more than they do the space program, but a plurality are satisfied with current priorities. Another immediate conclusion is that space means different things to different people. These observations are rather simplistic but perhaps accurate conclusions that can be drawn from responses to a single question in the General Social Survey (GSS), arguably the highest-quality long-duration scientific opinion research study carried out in the United States, which will be the focus of this chapter and contribute to later chapters as well. Before we can examine in depth the meaning of this single questionnaire item, it is worthwhile considering methodological issues that go beyond those discussed in the first chapter of this book, to understand the scientific power of the GSS.

Like many non-academic opinion polls, the General Social Survey seeks to use samples of respondents that represent the general population. One reason is to be able to describe the social conditions experienced by all Americans, projecting from a reasonably well-constructed and sufficiently large sample to minimize random errors that would be introduced by any bias in recruitment of people to answer the questions. Another reason is a well-established but frankly debatable view among social scientists that this is the best practical way to ensure that the relationships between variables are not the spurious result of unmeasured effects. A third reason that has deep intellectual consequences is the democratic value that each person's opinion should count equally in American society. Setting aside counter-arguments, the GSS provides a clear benchmark to contrast with data acquired by other means based on other methodological approaches.

W.S. Bainbridge, *The Meaning and Value of Spaceflight*,
Space and Society, DOI 10.1007/978-3-319-07878-6_3

3.1 The General Social Survey

There exist many kinds of important public opinion surveys, including those carried out by political campaigns, those conducted by commercial polling firms like Gallup and Harris, and those that have a more academic focus like the General Social Survey. When I was a visiting associate professor at Harvard University for 5 years in the 1980s, I came to know James A. Davis, the man most responsible for creating the GSS. I can remember discussing its goals with him at the dormitory called Winthrop House, of which he was Master, which meant he lived in a splendid apartment surrounded by student apartments. His life was embedded in academia, and he was committed to the great educational potential in the GSS. But he was frankly skeptical about over-arching sociological theories, and saw the GSS as a tool for charting changes in the conditions of life experienced by American citizens, which could be understood best by including opinion questions among those about more practical matters. Over the years, he and the rest of the governing board for the GSS became increasingly open to the inclusion of topical modules which could be carefully designed to test theories as well as to document social conditions. Today, Wikipedia describes the GSS thus:

> The General Social Survey (GSS) is a sociological survey used to collect data on demographic characteristics and attitudes of residents of the United States. The survey is conducted face-to-face with an in-person interview by the National Opinion Research Center at the University of Chicago, of a randomly selected sample of adults (18+) who are not institutionalized. The survey was conducted every year from 1972 to 1994 (except in 1979, 1981, and 1992). Since 1994, it has been conducted every other year. The survey takes about 90 min to administer. As of 2010 28 national samples with 55,087 respondents and 5,417 variables had been collected. The data collected about this survey includes both demographic information and respondent's opinions on matters ranging from government spending to the state of race relations to the existence and nature of God.

The question about the space exploration program was one of those about government spending, which Wikipedia chose to highlight and indeed have huge social scientific value (Kenneth and Rasinski 1989; Wlezian 1995). Its name in the GSS codebook is *natspac*, and the other questions about government funding all have names beginning with *nat*, presumably meaning *national* programs. In Chap. 5, we shall return to natspac to consider changes in support over the years, and it was asked in every one of the surveys except the first one in 1972. But here and in other chapters, the great value of the item is in its relationship to some of those other thousands of questions. In the introductory paragraph above, we focused just on responses from the latest GSS at the time of this writing, the one from 2012, and *year* is indeed one of those other variables.

One problem with the 2012 data is that only a subset of the complete group of respondents were asked the question, and a few of them failed to give clear answers. That year the GSS had a total of 1,974, only about half of them were asked this question, and the percentages reported above were based on exactly 899 of them. Other things being equal, the smaller the number of respondents, the less reliable the results are. Therefore, it is often wise to combine data across several years, thereby increasing the pool of respondents.

Combining all years gives us fully 31,227 respondents who answered natspac. Of this huge respondent pool, only 12.6 % wanted space funding increased; 40.5 % were content for it to stay at the level that was current in the year the person responded, and fully 46.9 % wanted space funding reduced. A small part of the difference between 2012 and all the years combined might be attributed to random factors in the smaller number who answered in 2012, but more substantively important would be the changing faction of the federal budget invested in space exploration, and evolution in the public's valuation of it. In roughly the first half of the period covered by the GSS, all the respondents received the question, whereas more recently only half did, so using this full sample of respondents overemphasizes the early years.

We can balance the reliability of a larger sample against relative stability in public opinion by combining just the surveys done in the years 2000–2012. That gives us a large number of respondents, 8,077, 15.3 % of whom wanted space funding increased; 45.6 % who wanted it kept the same, and 39.1 % who wanted it reduced. For much of this chapter, we will use this group, which facilitates comparing results across analyses.

One of the most significant explanatory variables, indeed one of the very few for which we have data about each and every respondent, is gender, called *sex* in the GSS codebook. Among men who responded in 2000–2012, 20.5 % wanted funding increased, 46.3 % wanted it kept the same, and 33.2 % wanted it reduced. Among women, the fraction wanting funding kept the same is very similar to that among the men, 45.1 %. But among women, only 10.8 % wanted funding increased for the space exploration program, versus 44.0 % who wanted it reduced.

I find the percentage wanting funding increased to be the most meaningful of the three responses, because it identifies people who have a clear perspective on the value of program, whether their views are factually correct or not, and identifying the people who have this opinion allows us to learn more about the factors that cause people to value space exploration. The often larger faction who want funding decreased may include a few thoughtful individuals who have analyzed the potential of the program and found it wanting, but more often probably people who have other priorities for federal funding. Those competing priorities are worth considering, but not so central to the theme of this study. The group who want funding kept the same may include a lot of people who just "don't know" and may not have thought about the issue. Thus, while the percentages for men and women overlap to a great extent, the difference between 20.5 and 10.8 % is quite noteworthy, and marks gender as a variable of prime concern.

Many of the other variables considered in this chapter correlate with gender, in some degree or another, and may help explain why gender has the effect it does. Sophisticated statistical analysis in the social sciences often uses techniques like multiple regression, path analysis, and log-linear analysis to tease out cause from effect in a complex network of interacting variables. For the population at large, gender is a perfect example of an *independent variable*, what non-specialists often call a *cause*, because it was determined at the beginning of a person's life. If, as we shall find below, a person's occupation affects attitudes toward space, and occupations vary by gender, we can imagine a simple causal sequence sex --> occupation --> space attitude. Both sex and occupation are important, but play different roles in the development of attitudes. However, sex may have a direct impact upon space attitudes, as well as the indirect effect via occupation. Also, within each gender, occupation may have an influence, or it may not.

Thus the possibilities in a complex and statistically sophisticated attempt to determine how many potentially causal variable fit together, influencing each other and being influenced in turn, are almost endless. While we shall often consider possible interactions between variables here, this is not the place for the kind of elaborate mathematical analysis required to nail down the relationships. Indeed, social scientists often debate each other's causal models, and such debates would be a distraction in the context of this book.

However, there is another methodological issue that must be considered, one that often requires sophisticated statistical analysis, namely the sampling that produced the set of respondents. Among the 8,077 respondents we have just considered, 3,718 were men, and 4,349 were women. That is, 53.8 % of the sample was female. But according to the 2010 US census, just 50.8 % of the population was female (U.S. Census Bureau 2010). Women tend to outlive men, so when those who are under age 18 are removed, the percentage female increases, and the numbers in the GSS partly reflect how many elderly people are in the sample. Over the years, the GSS has used a number of methods to achieve as representative a sample as possible, but no such techniques are perfect.

The first and most costly method to achieve quality in the GSS was mentioned in its Wikipedia article: the GGS "is conducted face-to-face with an in-person interview." A trained interviewer actually goes to the respondent's home, and administers the questionnaire as a 90-min interview. Other polls use telephone, Internet, and mailing paper questions with a return envelope, but as mentioned in the first chapter, these methods have low response rates, such that the respondents are often far from a representative sample. Going to the respondent's home not only achieves a higher response rate, but encourages the respondent to be serious while answering the questions. However, the cost of visiting homes everywhere across the United States would be prohibitively expensive, so the GSS has used cluster sampling methods, adjusting them over time, such that respondents live in a relative small number of general locations, reachable by an interviewer who is in that area.

The second method is applied by the GSS staff or individual research after the data have been collected, weighting the results such than they reflect the known distribution in the general population for a few variables such as sex and age. For example, when I use what the GSS calls a *composite weighting* procedure, combining three specific weightings, the percent of males wanting space funding increased becomes 20.2, and that for females becomes 10.9. These are different from the 20.5 and 10.8 % reported above. Without weighting, the percent of both genders combined who wanted funding increased was 15.3, and after weighting it is again 15.3, which indicates that the GSS did a good job of handling gender in collecting the data, but other respondent characteristics were not perfectly sampled in the interviews. In fact, of the three weighting procedures combined into composite, only one really matters, because the other two handled problems with surveys before 2000, such as the oversampling of African Americans in 1982 and 1987 the better to measure their opinions. The one procedure that was significant was compensating for non-response biases in 2004–2012.

I generally avoid weighting procedures in this book, both because is difficult and distracting to describe them, and because there always is room to debate which exact procedures should be used. In the case of the General Social Survey, there is a very compelling reason for avoiding weighting and other especially demanding statistical procedures, because the reader can duplicate and extend the findings reported here, at zero cost except for some rather interesting labor. The data are freely available either for download or for rather flexible online analysis from the Computer-assisted Survey Methods Program at the University of California, Berkeley at sda.berkeley.edu, or from the National Opinion Research Center of the University of Chicago that actually created the GSS and administers the surveys, at www.norc.org.

The excellent scientists at Berkeley and NORC deserve all the credit, but I played a small supportive role in developing these resources myself, and thus I am in an especially good position to use them, just as anyone else in the world can. For much of the 1990s, I managed federal funding for the GSS through my role as director of the Sociology Program at the National Science Foundation. Discussions early in 1994 led to a proposal to NSF to launch what may have been the first sophisticated service allowing anybody to analyze an important social science dataset online, through a $137,808 grant to NORC, and Berkeley was centrally involved from the very beginning. I had just become the representative of NSF's Directorate for Social, Behavioral, and Economic Sciences on the Digital Library Initiative, a marvelous multidisciplinary effort I compare in importance with the space program, for example in that it funded the research that led to Google. The Sociology Program was able to encourage other programs related to the Directorate for Computer and Information Science and Engineering to contribute to that frankly rather small grant by computer science standards. Here is the formal abstract describing the project, from the current NSF website:

> This is a prototype Internet service for the General Social Survey employing NCSA Mosaic. The project will develop a system to provide enhanced access to survey data, using the General Social Survey for implementation of these integrated services, which will subsequently be extended to a variety of other survey data sets. These services will provide facilities for hypertext viewing and searching of complete survey documentation, customized and documented extracts from data sets, statistical analysis, and File Transfer Protocol delivery of full or extracted data sets. The General Social Survey is an ideal source of survey material to develop the system, because it is a highly diverse large dataset of complex structure, extensively documented in terms of publications based on each item, and has already been the basis of more than three thousand scientific publications and dissertations. The system developed on the General Social Survey will then become

> the standard not only for providing survey data over Internet, but also for research design and data collection in new surveys.
>
> This project is part of the Digital Library Initiative, adding questionnaire survey data to the kinds of information that can be managed effectively over Internet and contributing to the national information infrastructure. Not only is the General Social Survey widely used in scientific research, but it has proven to be an excellent teaching tool at both the graduate and undergraduate levels. Survey data are used extensively in government and commerce, so an advanced system for managing and distributing information of this kind will contribute to effective government and economic growth.

"NCSA Mosaic" refers to the pioneering web browser created at the National Center for Supercomputing Applications, funded by NSF, which was the predecessor of Netscape and then of the more advanced web browsers used today. In that period I was myself migrating from Mosaic to Netscape, and served as the first webmaster for the single social science division that NSF then possessed. I am amazed to think how Internet and the Web have evolved over the two decades since the GSS online services were planned, yet they remain among the highest quality and most influential online social science resources, and will serve us well here.

3.2 The Other Problems Items

Groups of similar questions in survey research are often called *batteries*, and typically are introduced to the respondent with a brief explanation stating what the task of answering involves. This is what the interviewer would say before asking the questions about government funding: "We are faced with many problems in this country, none of which can be solved easily or inexpensively. I'm going to name some of these problems, and for each one I'd like you to tell me whether you think we're spending too much money on it, too little money, or about the right amount." In early years, the list included eleven questions, but by the year 2000 it had been expanded to fifteen, with natspac always being the first one.

It may seem strange to call the "space exploration program" a *problem*, rather than for example a *challenge*, but the sociological framework behind the battery of items was *social problems*. This is the title of many college classes, textbooks, and even a scientific journal. One of the fifteen items even includes the word *problem*: "solving the problems of the big cities." Frankly, I suspect the reason the space exploration program was listed first was precisely because it did not fit the problem definition well, and thus could help the respondent feel comfortable talking about the real problems that followed. Table 3.1 indicates how natspac fits with the other fourteen *nat* items.

The first two items in this table can help make sense of it more generally. Improving and protecting the environment is in some modest or indirect way connected in people's minds with the space exploration program. Among people who feel too little is being spent on space, 68.6 % feel too little also is being spent

Table 3.1 Percent feeling too little is being invested in each area, by space attitude

Government program	All, funding for space exploration is:		Males, funding for space exploration is:		Females, funding for space exploration is:	
	Too little (%)	Too much (%)	Too little (%)	Too much (%)	Too little (%)	Too much (%)
Improving and protecting the environment (natenvir)	68.6	61.4	68.0	59.0	69.5	62.9
Improving and protecting the nation's health (natheal)	65.9	76.9	64.5	71.4	68.2	80.4
Solving the problems of the big cities (natcity)	42.3	49.9	39.9	46.2	46.1	52.3
Halting the rising crime rate (natcrime)	49.4	65.9	46.6	59.7	54.0	69.9
Dealing with drug addiction (natdrug)	55.3	63.6	52.6	61.5	59.7	65.0
Improving the nation's education system (nateduc)	74.8	74.7	71.3	70.2	80.6	77.5
Improving the conditions of Blacks (natrace)	35.5	42.3	31.9	38.8	41.3	44.7
The military, armaments and defense (natarms)	33.6	25.5	31.1	24.0	37.6	26.6
Foreign aid (nataid)	13.0	9.2	12.8	9.3	13.4	9.1
Welfare (natfare)	23.4	27.0	20.9	25.7	27.5	27.8
Highways and bridges (natroad)	49.3	37.9	51.7	40.6	45.4	36.2
Social security (natsoc)	57.2	67.4	52.0	64.2	65.5	69.4
Mass transportation (natmass)	52.8	38.2	54.8	40.1	48.1	37.0
Parks and recreation (natpark)	38.4	36.0	37.6	37.7	39.7	35.0

on environmental programs. Among people who feel too much is spent on space, a modestly but statistically significantly lower fraction, 61.4 %, feel too little is spent on the environment. Perhaps the space program contributes to environmental protection in some way, possibly by monitoring the condition of the Earth from above. Or people may not connect the two directly in their minds, but some of the same kinds of people may support both. Perhaps these are people who like government programs in general, and do not mind the taxes they need to pay to support them all.

The item about improving and protecting the nation's health shows a very different connection to space support—a negative one. People who feel too little is spent on space are less likely than people who believe too much is spent to feel too little is spent on public health, 65.7 % versus 76.9 %. As a budgetary matter, all fifteen programs reflected in the table are in competition with each other for government funding, and for the taxpayer's dollar. Thus we would expect a negative correlation to be built into all the calculations, which would make any positive connections with the space program even more significant than they superficially appear.

In terms of public support for increased funding, six of the government programs are positively correlated with space exploration: environment, military, foreign aid, highways, mass transportation and parks. Seven of the programs are negatively correlated: health, problems of big cities, crime, drug addiction, the conditions of Blacks, welfare, and social security. Education does not seem to have either a positive or negative connection. There are many ways to characterize the government programs. But the positively correlated ones seem to be *progressive*, in the sense of building something new, often through the use of technology. The negatively correlated programs are *reactive*, collective responses to social problems, cures for misfortune that in some sense restore people to what we consider are normal conditions.

While this book will avoid getting mired down in multivariate statistical analysis and controlling for external variables in calculating correlations, the columns in Table 3.1 that report percentages for males and females separately are easy to understand and often informative. Men support the space program somewhat more than women do, and women support heath programs more than men do. But within each gender, these two programs show a comparable negative relationship. Gender apparently influences each of the attitudes, but the relationship between the attitudes may be quite similar across genders. This harmony is not always the case. The table can be an endless source of debate and perplexity, because the topics are substantively important and their relationships are sometimes complex.

For example, we might generally imagine that men are more favorable toward the military than women are, because they more often serve in the armed forces and are more often associated with violent actions in civilian life. For the 8,356 people who expressed an opinion on the natarms question, however, 27.1 % of males felt current appropriations for the military were too little, compared with a very slightly higher 28.5 % of women. Within each gender, support for increased space funding is positively associated with support for increased military funding.

Indeed, the highest level of support for military funding is the 37.6 % among that distinctive minority of women who favor the space program. Naively, we might speculate that men are more ready to fight, but women are more anxious to be protected, an idea we shall reconsider in the final chapter.

With both sexes combined, there is no appreciable correlation between support for space exploration and support for improving education, which is most immediately perplexing because the space program is often presented in educational terms, and as a motivator for interest in science among children. The situation gets more complicated when the sexes are considered separately. There is a positive but very weak connection between space and education in each gender, but women are more supportive of increased education funding regardless of how they feel about space. On the one hand, this situation illustrates how variables can interact with each other to produce complex results. On the other hand, it put us on notice that the role of education in spaceflight attitudes may be a complex one. We shall see more evidence of this in later chapters, and consider education more comprehensively in the final chapter.

Another perplexing finding in the table is that support for space exploration correlates negatively with support for increased welfare funding only among the men, and among women there is no connection. The opposite is true for parks and recreation, because only among women does a positive attitude predict greater willingness to invest in parks. While each such anomaly may deserve its own analysis, it would be helpful for the purposes of this book to have a way to distill the essence of the meaning of spaceflight, if indeed it has just one value.

3.3 Factor Analysis

While percentages like those in Table 3.1 are revealing, social scientists often use rather more sophisticated methods, such as correlation coefficients. But this can be daunting when many variables are involved. A complete analysis would look at correlations between all the possible pairs of items in this set of 15, which means $(15 \times 14)/2 = 105$ coefficients. Thankfully, social statisticians have developed several good methods to achieve *data reduction*, simplifying the picture without losing much information. Factor analysis is perhaps the most traditional, and we shall often use it in this book. It begins with the correlation matrix, and works to find some small number of abstract dimensions that can essentially graph the relations in a conceptual space. For our purposes this method is very appropriate, because each of these dimensions is likely to represent a different human value or ideological viewpoint.

In practical terms, the main result of a factor analysis is a table of new coefficients, similar to correlations but often further from zero in the -1.00 to $+1.00$ range, giving the *loading* of each item on each factor. If we graph the factors, the loadings determine how far out along each of the dimensions each of the items should be placed. If there are two factors, then ordinary graph paper will suffice. Decades ago, I constructed a three-dimensional graph of a factor analysis, using wooden balls for the items and dowels to support them in the air, but the analysis actually had four dimensions, so I painted them different colors in a spectrum representing the fourth dimension. Physicists may debate how many dimensions the universe actually possesses, but it is hard for humans to perceive more than three. The socio-cultural world certainly possesses far more than three, although admittedly they may be metaphors that help use conceptualize complex realities, rather than concrete directions in space.

In 1979, the technical director of the GSS, Tom W. Smith, did a factor analysis on the government funding items that were in the survey at that time, and reported: "that the first principal component factor was a social/welfare group consisting of education, welfare, blacks, cities, the environment, and health, and that defense, foreign aid, and space formed a national security factor (Smith 1979)." His report focused on support for increased educational funding, so it is noteworthy that education and space exploration wound up in two different factors. There are many forms of factor analysis, but most commonly the factors are orthogonal—at right angles to each other if displayed on a graph—which means there would be zero correlation between them and they represent totally independent aspects of public opinion.

Among the decisions that must be made when using statistical analysis software to produce factors is whether a specific number of factors should result. If we have a theory that predicts just two factors, we can do a *confirmatory factor analysis* that results in just two factors. This is not a sly way of forcing the computer to prove our theory correct, because the next step is to examine the factors and see if the items loaded on each one are those we predicted, and whether all the items are clearly associated with just one or the other factor. In this book I prefer the other approach, *exploratory factor analysis*, which does not specify how many must result, but does tell the computer to focus only on factors that explain a significant fraction of the total variation in the data. Technically, a common standard is all factors with *eigenvalues* greater than 1.

If the variables were not selected with any theory in mind, for example these government programs about which public opinion may vary, some resultant factors tend to group together many items, while others group few. In extreme cases, with many variables, the last factor or two may be rather meaningless, having no heavily loaded items. Sometime an item winds up significantly loaded on two factors, which only means that it has aspects relevant to at least two more general public debates.

We shall use exploratory factor analysis often in later chapters, but here it is especially useful for analyzing the fundamental meaning for public opinion of space funding, in comparison with the other government programs. Using all the government programs in the 2000–2012 data, and the 648 respondents who expressed opinions about all of them, six factors emerged, assigning each item to the one factor on which it had the strongest loading:

Factor 1: 0.75 Improving and protecting the environment
0.57 Developing alternate energy sources
0.55 Improving the nation's education system
0.53 Improving and protecting the nation's health
−0.60 The military, armaments and defense

Factor 2: 0.73 Welfare
0.66 Foreign aid

0.59 Improving the conditions of Blacks
0.42 Assistance for childcare

Factor 3: 0.68 Supporting scientific research
0.60 Highways and bridges
0.58 Mass transportation
0.53 The space exploration program

Factor 4: 0.71 Halting the rising crime rate
0.72 Dealing with drug addiction
0.33 Solving the problems of the big cities

Factor 5: 0.74 Social Security

Factor 6: 0.79 Parks and recreation.

The first factor is related to progress in science and technology, but seems *reactive*, or at best protective. However, "the military, armaments and defense" has a strong negative loading, −0.60, so this factor is certainly not an expression of totally general desires to be protected. I interpret this to means that respondents conceptualize the military in terms of technology, but those who favor environmentalism and the other positively-loaded items evaluate the military in negative terms. Another point of interest is the fact that the space program does not belong to this factor. Its loading on Factor 1 is −0.18, negative but very weak. The second factor could be described as *benevolent*, giving things of value to needy people. Both of these factors might be described as politically liberal, but representing very different sets of values, which people from a range of political viewpoints might favor.

Factor 3 includes "the space exploration" program and is led by "supporting scientific research." Exactly why "highways and bridges" or "mass transportation" are in this factor is unclear, but two compatible ideas come to mind. First of all, the four items involve progress, but of a different kind from that expressed in Factor 1, captured in the metaphor of *advancing* in a voyage of discovery. Spaceflight, after all, is a means of transportation, and a bridge to the planets. Second, all four items describe government investments in infrastructure, which are not politically controversial to any great degree.

Factor 4 emphasizes government's *policing* function, with essentially identical loadings for crime and drug addiction. "Solving the problems of the big cities" is also loaded 0.32 on Factor 2, and "the military, armaments and defense" is loaded 0.39 on Factor 4. Big cities have many problems which government addresses in two ways: imposing social order through police and related agencies, and providing welfare through programs to assist disadvantaged citizens. Respondents differ in which of these two conceptualizations they prefer, so the big cities item goes into two factors. The positive loading of the military reflects respondents who positively value its protective function, as armies can serve the same function as police, but on an international scale.

Factor 5 consists primarily of "Social Security," reflecting the interests of older respondents, and the fact that this government program is not conceptualized as a "hand out," which would have put it in Factor 2. The space exploration program is loaded −0.36 on Factor 5, indicating a tension with Social Security, and reflecting the low enthusiasm for space research among elderly people.

Factor 6 is primarily "parks and recreation," which may mean different things to urbanites versus suburbanites versus rural citizens. Interestingly, "the space exploration program" has a loading of 0.39 on this last factor, and "the military, armaments and defense" has a loading of 0.32, not large numbers but worth contemplating. Perhaps these three items express a *pioneering* instinct, if parks are a convenient way of exploring the wilderness, spaceflight enters the most vast possible wilderness, and the term *expedition* is often used in military terminology to describe an advance into dangerous territory.

By the general standards of factor analysis, this one is quite clear, although a little bit of imagination is required to name factors other than the one that is really just about the distinctive Social Security program. Social science has yet to develop a comprehensive theory of social attitudes, that would allow us to predict which ones are destined to correlate positively or negatively, or why gender differences occasionally reveal complex patterns. Part of the answer, and one of the reasons we have not yet arrived at a general theory, is because other factors have their own influence. We tend to call gender a demographic variable, because demographers and census takers count how many males and females there are in society, and two others that are often counted are age and race.

3.4 Demographics

Age can be considered from the standpoint of two rather different questions: In what year were you born? How long ago were you born? The birth year of an individual defines the span of historical events that person will experience. The number of years a person has lived measures where in the human life cycle the person currently stands. Table 3.2 shows the distributions of responses to natspac by decadal age groups, for a total of 8,058 respondents who answered in the 2000–2012 surveys.

From the 50–59 age group across the older groups, the percent feeling too little is being spent declines substantially, from 18.4 to 8.1 %. For those age 80 and

Table 3.2 Attitudes about space funding by age group

Age	Cases	Too little (%)	About right (%)	Too much (%)
18–29	1,418	15.2	50.4	34.5
30–39	1,605	14.9	46.9	38.2
40–49	1,669	16.5	45.4	38.1
50–59	1,419	18.4	42.2	39.4
60–69	982	15.7	45.2	39.1
70–79	621	9.3	44.9	45.7
80+	344	8.1	38.7	53.2

over, the majority want space funding decreased. Several factors could be at work. Older people might place higher value on social programs, and indeed Social Security is one of the government funding items. Some older people might favor space exploration in the abstract, but feel they themselves will not be alive to benefit from any more of its discoveries. Or perhaps their attitudes consolidated much earlier in life, before the success of the Apollo Program that reached the Moon in 1969.

This last possibility is what sociologist call an *age versus cohort* analysis. The people born in a particular period are a *birth cohort*, whose formative years were under roughly the same historical circumstances. Age concerns the point in the life cycle a person has reached. Notice that the people in the 50–59 age cohort are slightly more enthusiastic about space exploration than the younger groups, who went through their formative years after the great public space excitement of the 1960s. We will consider this age versus cohort issue more closely in Chap. 5, in the context of public attitudes about events that occurred during the history of astronautics.

A commonly discussed but conceptually antique demographic variable that strongly predicts some attitudes is race. The GSS asked every respondent, "What race do you consider yourself?" I know many people, including members of my own family, who would have difficulty deciding on an answer, or who might be offended by this very question. I indeed once refused to answer on behalf of my family the equivalent item on a US Census questionnaire. A representative of the Census Bureau even called me on the phone and coaxed me to answer, but I politely declined, suggesting the term *race* had no meaning, when I could have less politely asserted my family's privacy rights. But, given that most GSS respondents were comfortable answering the question, it is reasonable for us to look at the data they provided.

Among 6,242 "White" respondents, 16.8 % felt too little was being invested on space, compared with only 8.8 % of 1,142 "Black" respondents. In a miscellaneous category of 693 "other" respondents, 12.6 % felt too little was being spent. Much of the racial difference seems to reflect the fact that government programs are in competition against each other for taxpayers' money, and African Americans have ample reason to support programs designed to achieve greater social justice. This can be seen from how the categories of respondents differ in their attitudes about government investment for "improving the conditions of Blacks." Among 1,197 Black respondents, 75.6 % felt too little was being invested in the social programs designed to achieve this social goal. That contrasts with only 28.0 % of 5,939 White respondents who held the same view. As with the space item, members of the other racial category fell between those groups, 40.5 % of 640 feeling too little was being spent.

An ethnicity question that cross-cut the racial divide was, "Are you Spanish, Hispanic, or Latino/Latina?" I never knew what the Brazilian branch of my family should answer, because they are Latin American but not Spanish. In any case, 14.7 % of 791 Hispanics felt too little was being invested.

Place of residence is a standard demographic question, and the primary legal function of the US Census is to apportion political representation on the basis of the population of each area. It is not obvious why the place a person lives in the United States would affect attitudes toward space exploration, aside from a few communities whose economies depend on NASA facilities or industrial contractors, and given the GSS cluster sampling method we cannot use this dataset to look at those distinctive communities. Probably, any regional differences reflect the distribution of other variables, such as the big race difference we just identified, or more general economic conditions that vary across locations.

Of the nine standard census divisions of the United States, the smallest fraction who feel too little is being invested is the 11.8 % in the East South Central Region (Alabama, Kentucky, Mississippi, and Tennessee). I am well aware that NASA's Huntsville facility is in Alabama, having toured the place myself including the nice public museum of space history and technology. But world famous "Rocket City," which also calls itself the "Star of Alabama," may not heavily influence regional culture. The region with the highest percentage feeling too little is being spent is the Pacific at 18.3 %, just a hair higher than the 18.2 % for the Mountain region. Thus, the western part of the country is above the rest of the nation in terms of space enthusiasm.

One of the factors that distinguishes the western regions of the country is a somewhat high rate of geographic migration. The standard GSS question about this is: "When you were 16 years old, were you living in this same (city/town/county)?" Of those living in the same city, 13.8 % felt too little was being invested in space, compared with 15.3 % living in the same state but a different city, and 16.9 % of those living in a different state. Of course, migration reflects many social conditions, but it also may reflect a degree of personal risk-taking, exploration, or even adventurousness. Migrators may have some of the *Star Trek* spirit "to boldly go" to a new and different place, albeit one on our own planet.

3.5 Class and Occupation

Sociologists and many other social scientists place great emphasis on the variable of *social class*, although an objective definition of it has not been established. European social theorists, by no means limited to Marxists, have conceptualized society in terms of a class struggle, and their medieval history of feudalism dominated by a landed aristocracy gave them good reason to think that way. Many GSS items relate to social class, but two very distinctive ways to measure it are especially clear. One may define social class as a *subjective* condition, perhaps expressing people's general values and conception of their place in the world. Or one might define class as the *objective* positions people hold in the society's status structure, perhaps indicated by their occupation.

The GSS asks every respondent this question about subjective social class: "If you were asked to use one of four names for your social class, which would you say you belong in: the lower class, the working class, the middle class, or the upper class?" Table 3.3 shows the distributions of space attitudes across these four

Table 3.3 Attitudes about space funding by social class

Subjective social class	Cases	Too little (%)	About right (%)	Too much (%)
Lower class	535	11.4	36.4	52.1
Working class	3,584	13.8	43.0	43.2
Middle class	3,641	17.0	49.3	33.7
Upper class	277	18.8	51.6	29.6

groups, for a total of 8,037 respondents in 2000–2012. It is worth noting that two of the four classes, working and middle, have far more members than the other two. If social statuses were really in the shape of a pyramid, we might expect the lower class to be the biggest, rather than the second-smallest. Survey researchers generally recognize that "lower class" has a pejorative connotation, and many people whom social scientists would assign to the lower class prefer to call themselves members of the honorable working class. Those willing to call themselves lower class show the weakest support for space exploration, and a majority want funding reduced. Again, this may reflect their preference for social programs that might benefit them.

The social class differences are clear, but not as extreme as the gender differences, for example. This raises a profound question that will trouble us throughout this book: Given that the American public generally lacks great enthusiasm for increased space funding, does the program depend upon the enthusiasm of a social or intellectual elite, perhaps of professional scientists or of the so-called *power elite*. Here we see that the self-identified upper class is more enthusiastic than the middle class, but only slightly so.

Objective social class may be more meaningful than subjective, but is more difficult to measure. The GSS team has invested a great deal of effort in coding and analyzing people's occupations, so without trying to categorize them in terms of a simple class division, we can learn much for those data. For the 1988–2010 surveys, the jobs held by employed respondents were categorized in terms of major categories adapted by the GSS researchers from the 1980 US census occupation categories. Of the total 13,845 employed respondents, 13.3 % wanted to see space funding increased, but enthusiasm varied across occupational categories. Below are percents who feel space funding is too little, for all the categories to which at least 100 respondents belonged, adding together the farmers who specialized in plant crops and livestock animals to achieve 164 cases, so that the agricultural production category could be included:

3.8 % Apparel and accessories, except knit (131)
4.3 % Agricultural production (164)
4.6 % Beauty shops (108)
7.2 % Child day care services (111)
9.2 % Hotels and motels (130)
9.2 % Nursing and personal care facilities (240)

9.3 % Elementary and secondary schools (906)
9.6 % Department stores (228)
9.6 % Banking (218)
9.7 % Social services, n.e.c. (155)
9.8 % Private households (224)
11.1 % Hospitals (620)
11.2 % Motor vehicles and motor vehicle equipment (169)
11.3 % Eating and drinking places (688)
11.4 % Offices of physicians (114)
11.5 % Health services, n.e.c. (208)
11.9 % Trucking service (201)
12.9 % U.S. Postal Service (201)
12.9 % Justice, public order, and safety (248)
13.3 % Grocery stores (256)
13.5 % Insurance (222)
13.5 % Automotive repair shops (104)
14.3 % General government, n.e.c. (175)
16.0 % Business management and consulting services (100)
16.5 % Telephone (wire and radio) (164)
16.5 % Legal services (127)
16.7 % Construction (825)
16.7 % Motor vehicle dealers (102)
16.7 % Colleges and universities (354)
17.9 % Miscellaneous entertainment and recreation services (123)
18.5 % Electrical machinery, equipment, and supplies, n.e.c. (102)
18.8 % Printing, publishing, and allied industries, except newspapers (138)
19.9 % National security and international affairs (226)
20.9 % Business services, n.e.c. (134)
22.0 % Computer and data processing services (141).

These 35 job categories are arranged in ascending order of support for funding, and the number of people in each is given in parentheses. To be sure, these categories are only very roughly defined, and three of them are residual "n.e.c." (not elsewhere classified) groupings. Clearly, the list begins with low-education manual jobs, and ends with high-education and mostly technical jobs. This suggests that high-tech people are more favorable about that symbol of high technology, the space program. Many of them logically would see parallels between their own work and that of the astronautical engineers. Near the middle of the list we see several medical occupations, which often depend upon science-oriented expertise, but not of a sort that might be call "rocket science."

At 16.7 %, people employed at colleges and universities are much more favorable than those working for elementary and secondary schools, at a sub-average 9.3 %. Indeed, it is remarkable that support is so weak among employees of elementary and secondary schools, given that much pro-space rhetoric suggests that the space program inspires children to achieve academically. Of course, some obvious variables are involved. Females outnumber males in both groups. But are 77.0 % of the employees of elementary and secondary schools, versus 53.1 % of those working for colleges and universities. About equal fractions have had more than 16 years of education, 39.9 % versus 44.3 %, but the difference is greater for 20 or more years of education, just 7.1 of school employees having achieved this, versus 18.9 % of college employees. To be sure, many of these employees of educational institutions are not themselves teachers, yet one might have imagined that a general academic culture would have consolidated at all levels of education, favoring investment in space exploration as a crucial part of the life of the mind.

Many GSS surveys asked respondents about their service in the military, except for a wide gap in the period 1996–2008. At first glance, this variable appears very closely connected to enthusiasm for spaceflight, because only 10.9 % of 11,854 respondents who had never served wanted funding increased, versus 21.0 % of those 604 who had served more than 4 years. Yet military service here is a proxy for gender, and the differences narrow when only male respondents are considered. The numbers were 18.5 % for 3,970 men who never served, and 21.7 % for 580 who served more than 4 years. In 1982 and 1984, the GSS asked: "Have any other members of your immediate family—that is, your (spouse/parent/children/brothers or sisters)—ever served in the armed forces?" Of those answering, 1,547 said yes, and 629 said no. There was no difference in their support for space funding, indicating that differences by military service may reflect the characteristics of the individual, not the individual's family.

Looking across the three main military services, we see significant differences. Of 300 respondents who had served in the US Air Force, fully 28.3 % wanted funding increased, versus 22.6 % of 394 who served in the US Navy, and only 13.8 % of 1,039 who served in the Army. The GSS tabulated the Marines separately from the Navy, and 24.1 % of 116 respondents wanted funding increased. Narrowing the analysis to male respondents does not significantly change the results. Among 284 men who had served in the Air Force, 28.5 % wanted funding increased, versus 13.8 % again of exactly 1,000 who served in the Army.

Most obviously, support is greater in the service based on technology most similar to that of the space program, namely aircraft which have similarities to spacecraft, and to a lesser extent sea-going ships that may in some minds resemble spaceships. It is worth remembering that the original seven Mercury astronauts had been military pilots. Gus Grissom, Gordon Cooper, and Deke Slayton served in the Air Force. Scott Carpenter and Wally Shirra had been in the Navy, while Alan Shepherd and John Glenn were in the Marines.

3.6 Education and Ideology

A standard measure of social class is level of education, but schooling also shapes and reflects personal ideology. It can also be argued that people with more education tend to understand the realities of space technology better and be relatively well-informed

Table 3.4 Attitudes about space funding by educational attainment

Educational attainment	Cases	Too little (%)	About right (%)	Too much (%)
Less than high school graduation	1,119	8.9	37.8	53.3
High school graduate	4,118	13.9	44.9	41.3
Junior college	677	17.6	44.5	38.0
Bachelor's degree	1,407	19.8	51.2	29.1
Graduate degree	747	22.2	52.5	25.3

Table 3.5 Attitudes about space funding by scores on the GSS wordsum test

Test Score	Cases	Too little	About right	Too much
0–4	708	12.6	38.6	48.9
5–6	1,467	14.0	45.7	40.3
7–8	1,047	19.0	48.1	32.9
9–10	436	29.6	45.0	25.5

about its accomplishments and potential. Returning to the 2000–2012 data, 8,068 respondents told interviewers their level of education, analyzed in Table 3.4.

As was the case for relatively fine-grained occupational groups, the variations in levels of enthusiasm for space funding are rather large, and again the educational category associated with the lower class, those who did not graduate from high school, has a majority favoring reduction in funding. A cynical view of higher education is that it is merely a badge of higher social class background. But clearly, more educated people have greater knowledge and enhanced skills, at least in some areas of human endeavor. As a practical matter, more of them will have occupations that harmonize with the kind of technological development associated with the space program, however indirectly.

A very different way to measure education - or, forgive the term, intelligence— is through some kind of mental test. The GSS includes a ten-term word definition test, resulting in a *wordsum* score of correct answers, ranging from 0 to 10 (Cor et al. 2012). The GSS codebook does not list the words in the test, or the definition choices for each, in order to prevent respondents from studying it prior to their GSS interviews. A number of blogs assert that it is the same list as that used in a study of modern journalism posted online as a working paper (Sherr 2005). I cannot confirm that surmise, but the first word in that study's list happens to be *space*, for which the correct answer is *room*, thus not having astronomical connotations.

About half of the respondents answering the natspac item in the 2000–2012 period were given the test, and 3,658 people completed it. Table 3.5 shows there were substantial differences in space attitudes showing that greater verbal intelligence predicted stronger support. It is noteworthy that among those making only one error on the word test, or no errors at all, the faction feeling too little was being spent outnumbers those feeling too much was, 29.6 to 25.5 %. In the GSS data from this period, it is rare to see a group in which "too little" responses outnumber "too much" responses concerning the space exploration program.

Intellectual perspective is not just a matter of schooling, but also involves political ideology. One of the most widely used GSS items, employed by political

Table 3.6 Attitudes about space funding by political views

Political views	The space exploration program				Education system	
	Cases	Too little (%)	About right (%)	Too much (%)	Cases	Too little (%)
Extremely liberal	265	22.6	32.5	44.9	278	82.4
Liberal	790	18.2	49.2	32.5	843	79.4
Slightly liberal	742	17.3	49.2	33.6	789	77.8
Moderate	2,504	13.7	44.9	41.4	2,658	75.4
Slightly conservative	954	15.0	45.2	39.8	1,001	69.8
Conservative	1,014	15.6	48.2	36.2	1,062	62.1
Extremely conservative	251	18.3	37.1	44.6	260	55.0

scientists as well as sociologists, is *polviews*: "We hear a lot of talk these days about liberals and conservatives. I'm going to show you a seven-point scale on which the political views that people might hold are arranged from extremely liberal—point 1—to extremely conservative—point 7. Where would you place yourself on this scale?" Then the interviewer would hand the respondent a card showing the seven categories listed in Table 3.6, which has statistics based on 6,520 respondents who answered both polviews and natspac.

Remarkably, the percents feeling funding was "too little" are highest at the two political extremes, and lowest among moderates. Conceivably, Liberals like scientific discovery, and Conservatives like the national strength reflected in the traditional symbiosis between space and military technologies. Or, the political extremes attach two other very different meanings to space exploration that connect it to their basic political values. Or, it could be that the extremes are simply more ready to express strong opinions and moderates suffer from shyness. Are moderates really "middle-of-the-road," or are they often what Robert K. Merton called *retreatists*, pessimistic people who have given up hope that conditions can be improved? (Merton 1968) Some of them may simply lack opinions, but note that the fraction saying that space funding should be kept the same is smaller than in any of the other groups except the two extremes. This would indicate that many of these moderates really want reduced funding. Thus, it seems likely that space exploration does have a different positive meaning for political Liberals versus Conservatives.

To provide an efficient comparison, Table 5.6 includes two columns for the item that asked about funding for "improving the nation's education system." Here the pattern is very different, greatest support among Liberals and smoothly declining support across the spectrum to the Conservatives. Moderates are in the middle, where logically they ought to be if they are forthrightly expressing their views rather than being captives of diffidence. This is one more way in which the connection between education and space exploration is not what we might simplistically imagine it to be.

Table 3.7 Attitudes about space funding by political affiliation

Political position	Cases	Too little (%)	About right (%)	Too much (%)	College graduate (%)	Male
Strong Democrat	1,317	12.6	41.5	45.9	28.8	37.4
Not strong democrat	1,413	14.4	41.8	43.7	25.7	41.3
Independent, near democrat	949	17.6	45.6	36.8	28.8	49.6
Independent	1,429	13.4	43.3	43.3	14.9	45.6
Independent near republican	688	19.6	49.7	30.7	30.5	56.1
Not strong republican	1,193	14.4	51.7	33.9	31.1	49.3
Strong republican	897	17.4	51.2	31.4	31.4	48.7
Other party	154	26.0	42.2	31.8	35.1	61.7

To complete our analysis of education and ideology, and discover that the complexities are quite considerable, we can introduce a GSS item about political party affiliation: "Generally speaking, do you usually think of yourself as a Republican, Democrat, Independent, or what?" People who responded "Republican" were then asked, "Would you call yourself a strong Republican or not a very strong Republican?" Democrats responded to a similar question, and Independents were asked, "Do you think of yourself as closer to the Republican or Democratic Party?" Table 3.7 shows the result of analyzing data from the 8,040 respondents who answered this set of questions plus the space funding items in 2000–2012. The columns giving the percents college graduate and male are based on this same subset of respondents.

The data were collected during a period when the space program was not a salient political issue, as it may have been during the Kennedy-Johnson and Reagan administrations, and the table shows very complex patterns. Strikingly, ordinary Democrats and Republicans show exactly the same levels of positive support, 14.4 % saying that current funding is too little. However, they differ greatly in the fraction feeling that too much is being invested in the space program, 43.7 % among not-strong Democrats versus only 33.9 % among not-strong Republicans. Opposition to space funding is weaker among Democrat-leaning Independents (36.8 %) and stronger among strong Democrats (45.9 %), suggesting that within the Democratic Party a key issue is that funds could be moved from space exploration to social programs. The differences across degrees of Republicanism, in contrast, are not remarkable.

The Independents who lean toward one or the other of the two parties show more interest in an expanded space program than do members of the parties, perhaps reflecting unusually active thinking about public issues, given that they are not willing to subordinate their personal opinions to the ideology of an organized party. Independents who do not lean toward either major party show the second-lowest fraction who feel space funding is too little, 13.4 %, and are evenly split between feeing funding is about right versus too much. Only 154 of the 8,040 respondents belonged to a political party other than the Democrats and Republicans. So the estimates of third-party attitudes in the general population are not very reliable, but fully 26.0 % of these respondents feel too little is being invested.

The column of Table 3.7 giving the percentage who were college graduates in each group is based on 8,032 respondents who fully answered the relevant questions, and indicates that Republicans on average are very slightly better educated than Democrats, which may just be a reflection of the social class interests expressed by the two competing parties. The most striking figure in that column is the mere 14.9 % of the pure Independents who have completed college. This group may simultaneously be less well informed than the others, but also may be relatively indifferent to public issues.

The final column of the table reveals that males are overrepresented among those who are party-leaning Independents or members of other parties. It is a standard finding of sociology, even a cliché in criminology, that males tend to violate societal norms more than females do. To the extent that space exploration is revolutionary, this propensity may partially explain the greater male enthusiasm for the space program seen in many sets of poll data. Consideration in the light of the education variable offers a more favorable interpretation. In a 1979 GSS working paper, James A. Davis saw "an element of progressivism" in the responses of educated people to many items including the one about space funding: "Educated people tend to give more support to the new: abortion, space exploration, and women's equality. To me, many of the items also convey a flavor of optimism, lesser rigidity, and lower hostility (Davis 1979)."

3.7 General Social Conclusions

By and large, the American public supports continued but not expanded space exploration. In these data from the General Social Survey, we find no evidence that cultural lag has vanished over the period 1973–2012. Nor do we see any special desire on the part of the public to be especially inspired by the accomplishments of the space program over this period (Delgado 2011). Demographic variables like gender and age have some power to shape attitudes, perhaps in that many women and elderly people may not see how spaceflight can benefit them, in meeting the challenges of their every-day lives.

The inclusion of the GSS item on space funding in a battery of similar items about other government programs reminds us that NASA competes with other agencies, and exploration competes with other needs. Factor analysis clarified how government programs cluster together in serving more general values. To the extent that space funding correlates positively with other programs, then funding them instead of space may serve their shared fundamental values perfectly well. To the extent that space funding correlates negatively with other programs, then their supporters implicitly become opponents of the space program.

Because it is not inexorably tied to one major political party or the other, space exploration is not a high priority for either Democrats or Republicans, and its recent inability to inspire the general public means they both are unlikely to give it a high priority on political grounds. However, the fact that a significant minority of people are enthusiastic about it, especially thoughtful Independents, a governing party would lose voters if it ended manned spaceflight. These enthusiasts seem especially common among the educated and technological elite, thus able to influence events to some degree outside the ordinary political process. This suggests we need to supplement our analysis of general public opinion with the views of a societal elite.

References

Cor, M. K., Haertel, E., Krosnick, J. A., & Malhotra, N. (2012). Improving ability measurement in surveys by following the principles of IRT: The wordsum vocabulary test in the general social survey. *Social Science Research, 41*, 1003–1016.

Davis, J. A. (1979). TR02 background variables and opinions in the 1972-77 NORC GSS—ten generalizations about age, education, occupational prestige, race, religion, and sex, and forty-nine opinion Items, Topical Report, Chicago, NORC, 8, p. 50. Retrieved December 8, 2013, from publicdata.norc.org:41000/gss/DOCUMENTS/REPORTS/Topical_Reports/TR02.

Delgado, L. M. (2011). When inspiration fails to inspire. *Space Policy, 27*, 94–98.

Kenneth, A., & Rasinski, K. A. (1989). The effect of question wording on public support for government spending. *The Public Opinion Quarterly, 53*(3), 388–394.

Merton, R. K. (1968). *Social theory and social structure*. New York: Free Press.

Sherr, S. (2005). News for a new generation: Can it be fun and functional? Eagleton Institute of Politics, Rutgers University. Retrieved October 27, 2013, from www.civicyouth.org/PopUps/WorkingPapers/WP29Sherr.pdf.

Smith, T. W. (1979). SC13 public support for educational spending, Social Change Report, Chicago, NORC, p. 10. Retrieved December 8, 2013, from publicdata.norc.org:41000/gss/documents//SCRT/SC13%20Public%20Support%20for%20Educational%20Spending.pdf.

U.S. Census Bureau. (2010). Age and sex composition, 2010, Census Brief, May 2011. Retrieved October 26, 2013, from www.census.gov/prod/cen2010/briefs/c2010br-03.pdf.

Wlezian, C. (1995). The public as thermostat: Dynamics of preferences for spending. *American Journal of Political Science, 39*, 981–100.

Chapter 4
Goals in Space

Expensive public opinion polls using respondents who represent the general population, such as the General Social Survey, have much to contribute to an understanding of what the space program means to the average citizen. But they are not well suited to explore the topic in depth, both because their high cost limits the number of questions that can be asked on any one topic, and because most respondents will lack the information or aptitude to analyze the full range of values space exploration may have. Thus there also is a role for studies that use specialized samples of respondents, and ask many intellectually demanding questions.

In January 1986, I happened to be doing observational research at NASA's Jet Propulsion Laboratory, looking at how scientists and members of the press were collaborating to find the right metaphors to communicate to the general public about the encounter of the space probe Voyager II with the planet Uranus. We all watched the launch of the Challenger space shuttle on NASA's direct television feed from Florida, because the networks had stopped carrying these events under the assumption they had become routine. The explosion of the Challenger inspired me to carry out a two-phase questionnaire study with students at Harvard University, to chart the distinct goals well-informed people believed might justify a continued space program. A fresh analysis of what are now historical data from that study will be the focus of this chapter, but first a conceptual and methodological framework must be established.

4.1 Ethnographic Questionnaires

The approach used in this study has some of the quality of *ethnographic field research* in cultural anthropology, or what is sometimes called *grounded theory field research* in sociology. As I have explained in my recent book, *Personality Capture and Emulation*, the process requires the researcher to communicate intensively with

W.S. Bainbridge, *The Meaning and Value of Spaceflight*,
Space and Society, DOI 10.1007/978-3-319-07878-6_4

members of the target population, using open-ended qualitative questions in interviews or a questionnaire, to develop the fixed-choice items for a later questionnaire, the final data from which would be analyzed statistically (Bainbridge 2014). Many social-science questionnaires seek to test pre-existing theories, which means that their items will be derived from the existing literature of the particular social science, or from the armchair ruminations of the researcher. Ethnographic questionnaires are in a sense written by the respondents who belong to the culture under study, with technical assistance from the researcher but expressing their views rather freely.

The 1986 study actually began in September 1972, when I conducted tape-recorded interviews with a non-random sample of 58 participants at the World Science Fiction Convention in Los Angeles. I asked them such questions as: "Why do you think we ought to go into space?" "What do you think man's ultimate future in space might be?" "Is there any aspect of spaceflight that interests you especially?" "What do you think the result would be if we stopped going into space altogether?" "Looking toward the far future, what do you think ought to be done in space?" These somewhat broad questions were meant simply to stimulate open-ended talk about the value of spaceflight.

Using the somewhat primitive information technology of the time, I went through the tape recordings, using a typewriter to put on 242 file cards the distinct answers the science fiction fans and authors had given. Following the methodology called *grounded theory*, I then sorted the cards into 28 piles, each expressing what seemed like a distinct clear idea (Glaser and Strauss 1967). In the methodology of grounded theory, each category should become a distinct statement as it becomes "saturated" by the addition of more and more contributing utterances.

Despite having invested years of effort in several research studies using qualitative methods, I would be the first to admit that ethnography tends to lack rigor, and non-quantitative observational research is rather subjective. Twin remedies can overcome these problems: (1) Reducing subjectivity by being as systematic as possible, and candid in reporting results; (2) Moving as soon as appropriate to quantitative methods, summarizing the judgments of many people whose biases are likely to be diverse. A diversity of subjective viewpoints not only balances each bias against several others, but provides the multiplicity of perspectives that can provide a more complete picture of reality.

In May 1972 I had begun observational field research on a private spaceflight-boosting social movement called the Committee for the Future (CFF). With the interview data from the science fiction convention to provide orientation, I scanned through CFF literature and tape recordings I had made at its SynCon convention, to begin assembling its ideology. I then mailed questionnaires to the people who had attended the CFF meeting, and to members of the New England Science Fiction Association (NESFA), including this question: "In your opinion, what is the most important reason why we should continue the space program?" NESFA had been founded in 1967, a spin-off from the MIT Science Fiction Society; both groups remain very active today, and are central to the science fiction subculture.

Science fiction fans and participants in the Committee for the Future could be described as visionaries, rather than scientists and engineers, so it was essential to expand the respondent sample to include some more technical people. An obvious choice was the membership of the preeminent engineering organization in the field, The American Institute of Aeronautics and Astronautics (AIAA). A questionnaire including the items from the NESFA and CFF survey, plus four more like it, was sent to a random national sample of AIAA members, and 102 of them responded (Bainbridge 1978). All the responses from all sources were typed out on file cards, one utterance per card, with code numbers identifying the source of each statement.

In the first two stages of analysis, I employed two practiced coders from the Harvard social science community where I was then a graduate student, Mrs. Pamela Smith and Ms. Pamela English. Both were well-trained in data analysis but unfamiliar with spaceflight ideology and uninformed about the sources of my data or goal of my research—thus unbiased. Smith did the first analysis, sorting the 1,260 cards into 63 different categories of justification for spaceflight, basing her classification on her own perception of similarities among the statements. Then, with that background but working independently, English did the second sorting, which I will describe here, as a logical step toward later stages in this research, but offering its own insights.

English was asked to go through the cards twice, without knowing which of the three groups had contributed each statement. First, she applied an abstract "near-far" 5-step scale to each card. If one seemed very near, she coded it "1," and very far was coded "5." The integers 2 through 4 indicated degrees of distance between the two extremes. I suggested to her that she use her own abstract, even intuitive concept of near versus far, rather than how near versus far in the future or in spatial distance each one might be situated. She said she had no difficulty doing this. Second, she assigned each card to one of seven categories developed on the basis of the work done by Smith and myself: Scientific, Technological, Economic, Political, Social, Psychological, or Religious. Table 4.1 shows the results of her work, for 1,256 cards because 4 could not be classified.

Table 4.1 Classification of 1,256 statements about spaceflight

Space program benefit category	Percent of responses in category				Average "far" score
	Science fiction fans	Committee for the future	American Institute of Aeronautics and Astronautics	Three groups combined	
Scientific	25.3 %	13.9 %	20.2 %	20.1 %	3.73
Technological	15.6 %	19.9 %	40.5 %	28.9 %	1.87
Economic	10.0 %	13.5 %	14.4 %	13.0 %	1.80
Political	12.4 %	15.5 %	16.5 %	14.9 %	2.81
Social	7.6 %	6.1 %	1.3 %	4.1 %	3.25
Psychological	22.1 %	11.8 %	5.6 %	11.5 %	3.70
Religious	7.1 %	19.3 %	2.1 %	7.5 %	4.37
Total	100.0 %	100.0 %	100.0 %	100.0 %	2.83
	(340)	(296)	(620)	(1,256)	

The three groups show different patterns of contribution across the seven categories. The most common categories of statements for the science fiction respondents are scientific and psychological; for the CFF the most populated categories are technological and religious, and for the AIAA, technological and scientific. Combining responses from the three groups, and looking at the near-far scale, economic and technological goals are the nearest, and religious goals the farthest. Of course differences between the groups partly reflect the different ways their data were collected, but the reasonableness of the results encouraged me to use the material as the basis of a more serious classification effort. With the classifications done by Smith and English in mind, I carefully sorted the 1,256 meaningful statements into groups, repeatedly combining, separating, and reassigning cards, until 49 categories had been saturated.

4.2 The Seattle Voter Study

In preparation for a new quantitative questionnaire, I wrote a sentence summarizing what I thought was the main idea in each of the 49 groups of statements. At times this felt like translating poetry into prose, or splitting hairs, but this seemed the best practical way of transitioning to the next stage in an increasingly rigorous and ambitious survey research program. To provide some hint of the process and the challenge, here are two paragraphs quoting text from some of the cards in a category, and beginning with my summary sentence, for two ideas that had a considerable degree of overlap, while also being rather visionary in nature:

> Our world has become too small for human civilization and for the human mind; we need the wide open spaces of the stars and planets to get away from the confines of our shrinking world. "We are quickly outgrowing our own world, and need space to spread." "Earth is now too small for human civilization and for the human mind. If it were penned for much longer within the narrow confines of earth, our civilization would evolve into something far more horrible than anything in a quite sufficiently horrible past." "Lebensraum" is needed "to... give the collective mind an 'out' - a place to go to be free." "You can't hold in everybody in a society. You can't fence the whole society in... There's people that just have to be let out. It's like the old west... Some people just have to live in the wide open spaces. And there isn't anything wider or opener than space." "There must be a place to escape from civilization." "There are always individuals who would rather not live with the civilization they find forming around them... and they are the ones who are going to strive to pass the astronaut exams, because they'll eventually be about the only frontiers left... They have to go outward bound."
>
> Overpopulation on Earth can be solved by using the living space on other planets. "Factors on earth may compel colonization of other planets." Because of "overpopulation and continued war," "we're going to have to find somewhere to put people in the future." "We are either going to choke to death or starve to death or maybe get squeezed to death on this planet because the planet has grown too small." "Overpopulation will reach a point where space travel will be an absolute necessity... The only answer would be space travel to put people on other worlds." "We need more room!"

In 1977, after I had settled in at the University of Washington, Richard Wyckoff and I mailed a survey based on the 49 spaceflight goals to a random sample of registered voters in Seattle, receiving analyzable replies from 225 voters, representing a response rate of about 45 % (Bainbridge and Wyckoff 1979). While not a high response rate, this probably included most of the invited voters who had well-developed ideas about space exploration, and I proceeded to use factor analysis to find the more general values that were expressed by groups of items. Respondents were asked to judge how good a reason for supporting the space program each of the 49 goals was, using a 4-point scale: not a good reason, slightly good reason, moderately good reason, and extremely good reason. Five main groups emerged, as shown in Table 4.2. I gave each factor a name that seemed to me to summarize its thrust, and calculated its *popularity*, defined as the mean percentage across the items calling each one "extremely good" or "moderately good."

The numbers of items pulled together into clusters range from 5 to 14, which reflects the fact that the general concept behind each cluster could be expressed through specific goals that were distinguished from each other in more or less fine degree. The Information factor lists ways that space contributes to the discovery and communication of new knowledge. It lists four already successful programs: weather satellites, navigation satellites, Earth resource satellites, and communication satellites. These systems collect and distribute information. The factor suggests that "Space technology will allow us to manage the environment of our planet because it is developing techniques for managing artificial environments that support human life." The key principle is information, whether of an immediately practical or more abstract nature: "Space development will give us new practical knowledge that can be used to improve human life." "Space exploration adds tremendously to our scientific knowledge."

The Economic-Industrial factor talks about the job opportunities and economic stimulus provided by the space program. It says we must continue the program in order to maintain the quality of American technology and so that our highly trained manpower will not be wasted. It says the space program encourages young people to choose careers in science and technology, and that the program is a good training ground for scientists and engineers. Finally, the factor mentions spin-offs: "Space technology produces many valuable inventions and discoveries which have unexpected applications in industry or everyday life." This is the only statement shared by two factors, as defined by factor loadings greater than 0.40, also belonging to the first factor.

Within the Military factor, the most popular item was also the one with the weakest loading: "Military reconnaissance satellites (spy satellites) further the cause of peace by making secret preparations for war and sneak attacks almost impossible." Its loading was 0.56, while the item about "military applications" has a loading of 0.77. The popularity of "reconnaissance satellites" was 59.8 %, compared with 44.0 % for "military applications." The Emotional-Idealistic factor mentions a number of personal feelings and spiritual motives that might be served by spaceflight. It says we must explore space to satisfy our great curiosity and in search of fun, excitement and adventure. Space provides a challenge and a goal for mankind, an outlet for human aggressive instincts, and may help bring about global renewal on Earth. Space enlarges the mind and the spirit of man, and will teach us to love and respect our own planet. The factor even includes the

Table 4.2 Factor analysis of 49 summary statements about spaceflight

Factor	Two mostly highly loaded items	Items	Popularity (%)
1. Information	Navigation satellites are a great help to ship and plane navigators, and traffic control from space can aid safe and efficient use of conventional transportation systems. Meteorology satellites aid in making accurate predictions of the weather	11	76.0
2. Economic-industrial	The space program provides an essential stimulus to the whole economy by investing money and paying employees. We must continue the space program in order to maintain the quality of American technology	6	48.0
3. Military	Space has military applications; our nation must develop space weapons for its own defense. Space is an important arena for international competition, and if we do not keep our lead, the Russians will gain an advantage over us	5	43.4
4. Colonization	Overpopulation on Earth can be solved by using the living space on other planets. Space travel will lead to the planting of human colonies on new worlds in space	14	33.3
5. Emotional-idealistic	We must explore space to satisfy our great curiosity; space exploration is an expression of man's natural curiosity. Space exploration is an exciting adventure, valuable for the fun and excitement it provides	13	32.5

following personal statement: "I am in favor of the space program because I would very much like the experience of traveling into space myself."

Table 4.3 lists all the items in the Colonization factor, because they suggest the possible far future of spaceflight in rather clear terms. The city in which the respondents lived, Seattle, was named after a Native American leader, yet was the result ultimately of European colonization, and is within the state of Washington, named for the Revolutionary War leader who became first president of the United States. At the time, the city was the home of the Boeing aerospace company, and it was a major conduit for trade with East Asia. The Microsoft corporation had not yet moved its headquarters to nearby Redmond, Washington, but Seattle already thought of itself as a high-tech town. Thus, the relative unpopularity of the colonization factor may set an upper bound for public enthusiasm as of 1977.

Table 4.3 The Seattle voter colonization factor

Statement	Loading	Popularity (%)
Overpopulation on Earth can be solved by using the living space on other planets	0.70	24.9
Space travel will lead to the planting of human colonies on new worlds in space	0.70	24.3
Society has a chance for a completely fresh start in space; new social forms and exciting new styles of life can be created on other worlds	0.66	24.0
Raw materials from the moon and other planets can supplement the dwindling natural resources of the Earth	0.63	50.9
Our world has become too small for human civilization and for the human mind; we need the wide open spaces of the stars and planets to get away from the confines of our shrinking world	0.59	17.6
Spaceflight is necessary to ensure the survival of the human race against destruction by natural or man-made disaster	0.57	25.6
Human societies have always needed to expand in order to remain healthy; space is the only direction left for such expansion	0.56	31.4
We must go beyond the finite Earth into infinite space in order to continue economic growth without limit	0.54	20.7
Space hospitals put into orbit where there is no gravity will be able to provide new kinds of medical treatment and give many patients easier recoveries	0.53	50.7
Commercial manufacturing can be done in space without polluting the Earth; completely new materials and products can be made in space	0.47	40.6
Communication with intelligent beings from other planets would give us completely new perceptions of humanity, new art, philosophy, and science	0.44	55.3
We can conduct certain dangerous kinds of scientific experiment far in space so accidents and other hazards will not harm anyone	0.42	36.2
Without spaceflight we would be trapped, closed-in, jailed on this planet	0.41	14.7
Rockets developed for spaceflight will be used for very rapid transportation of people, military equipment, or commercial goods over long distances on the Earth	0.38	49.1

This Colonization factor contains a wide range of ideas that had originally been proposed by the members of the three space-oriented groups, and thus has a radical quality. The least popular are the most obviously radical, because they fundamentally reject humanity's home, the Earth: "Our world has become too small for human civilization and for the human mind; we need the wide open spaces of the stars and planets to get away from the confines of our shrinking world." "Without spaceflight we would be trapped, closed-in, jailed on this planet." This is the thinking of a revolutionary social movement, more than of the typical citizenry.

4.3 The Harvard University Study

Nine years after the Seattle study, the opportunity and motivation for a replication on a much larger scale presented itself, by administering a pair of questionnaires to volunteers in the dining halls of Harvard University (Bainbridge 1991). In Phase 1 of the study, 1,007 students completed a questionnaire I call S1986A, including four open-ended questions asking them to express freely any ideas they had about the values served by space exploration: "In your opinion, what is the most important reason why we should continue the space program?" "Can you mention a very different benefit of the space program?" "Some perfectly valid and important justifications for the space program are often ignored and deserve greater mention than they commonly receive. Can you give us such a justification?" "Can you mention a possible long range result of a vigorous space program that would eventually be significant for humanity?"

This section of S1986A came right after eight items that measured respondents' level of enthusiasm for the space program, with room for them to write comments after checking their selected box, giving them the opportunity to express negative attitudes as freely as positive ones. Aware that some respondents might disagree with the positive framing of the four open-ended questions about the value of the program, I provided a disarming introduction: "Unlike the questions above, these seem to assume you have a favorable attitude toward the space program—but please do not be put off by this. The purpose of this section is to collect many ideas about why people might support the space program or various space projects. All these ideas will be sifted carefully, compared with each other, and incorporated in a future questionnaire that will assess the enthusiasm (or lack of enthusiasm) that Harvard students have toward each of them. Thus, we would greatly appreciate your responding to these questions, whether or not you personally support the space program."

As in the earlier pilot study, all the responses were written out and sorted into groups that seemed to express the same idea. Given the large number of respondents, their diverse interests, and their generally high awareness of the space program after the recent Challenger disaster, they contributed a large number of ideas, which totaled 125 once I had sorted them. As before, I wrote a summary statement for each of the resultant groups, drawing on the text giving the clearest expression.

In Phase 2, through questionnaire S1986B, another set of students rated each of the statements in terms of how good a justification it was for supporting the space program, on a scale from "0" (not a good reason) to "6" (extremely good reason).

The introduction on the first page of S1986B explained that the 125 items had been derived from an earlier questionnaire completed by 1,007 Harvard students, and included this disclaimer: "Whatever your feelings about the space program, we hope you will share them by completing this questionnaire. While most of the items in this questionnaire are stated in an apparently 'pro-space' manner, there is ample opportunity for you to express negative sentiments, if you have them." The instructions for the main section listing the 125 "justifications for the space program" added: "You will probably feel that some reasons are much better than others. Don't worry about all the aspects of each one, but make an over-all judgment of it." There were five versions of S1986B, differing only in what random order the goals were presented, a method intended to minimize correlations that merely reflected a tendency to give similar ratings to adjacent items.

Altogether, S1986B contained 147 questions, and my analysis used data from just the 894 Harvard students who skipped no more than 5 of them. A form of cluster analysis called *block modeling* was used to derive a smaller number of very general values potentially served by space development. There were two reasons I used the block modeling method of statistical analysis. First of all, as a practical matter, I did not then possess statistical software that could apply traditional factor analysis to a data matrix as large as $125 \times 125 = 15{,}625$ cells. So I programmed my own software from scratch. Second, I was interested in exploring the potential of block modeling, which had been developed by my former professor Harrison White, and which I conceptualized as an interesting variant on multi-dimensional scaling, a computationally-intensive new method that was then beginning to compete with factor analysis, but has subsequently faded from interest in sociology (White et al. 1976).

Returning now to the data after more than a quarter century it seems both methodologically appropriate, and frankly ironic, to see what the very traditional factor analysis method can derive from the data. The analysis that follows does not contradict the results reported in my book *Goals in Space*, but offers a very different perspective on them. At a 2003 conference I reported that factor analysis and block modelling gave complementary perspectives on the structure of this dataset, but this chapter is the first full report of the results of factor analysis (Bainbridge 2004).

The complex factor analysis presented here was performed with a subset of 90 items which proved especially meaningful, and a subset of 512 respondents, 256 males and 256 females who had responded more thoroughly than the others. The full set of 894 respondents will be used to determine how each of the 90 justifications for the space program connects to actual support, as measured by the space funding item from the General Social Survey, which was included in S1986B, an analysis that has not previously been reported in any form.

Factor analysis of data from ethnographic questionnaires often places many items in Factor 1, because they measure some very general influence on public opinion, and it can be useful to reanalyze those items to find additional meaning that is obscured in the original analysis. That proved to be the case with these data, so the factor analysis was done in two stages. First, all 90 items were subjected to an exploratory factor analysis (seeking all factors with eigenvalues greater than 1, and using varimax rotation). Then, all items that had loadings on Factor 1 of at least 0.40 were subjected to a second factor analysis of the same kind. The 0.40 threshold is a rule of thumb I have found to be reasonable in past research, although as will be mentioned below, some items have loadings this large on two factors, rather than "belonging" only to one factor.

The two-stage exploratory factor analysis indeed produced a very large Factor 1, and then separated it into four subfactors with very clear meanings: Subfactors 1A, 1B, 1C, and 1D. Following are the items in each, with an item's loading on the subfactor:

Subfactor 1A: Inspiration

- 0.81 The space program builds national pride.
- 0.78 Space triumphs give us justified pride in our achievements.
- 0.71 The exploration of space lifts morale and instills a sense of hope and optimism.
- 0.69 Spaceflight reaffirms faith in man's abilities.
- 0.61 Spaceflight is a noble endeavor, expressing the hopes and aspirations of humankind.
- 0.57 The space program provides a goal and a feeling of long-term purpose for humanity.
- 0.56 The space program encourages people to make achievements and solve problems.
- 0.54 Space exploration is a human struggle, expressing the unconquerable human spirit.
- 0.51 The space program allows people to think beyond the triviality of Earth-bound conflicts and concerns.
- 0.44 The space program inspires young people to study the sciences.

Subfactor 1B: Exploration

- 0.75 We should explore the unknown.
- 0.71 We should boldly go where no man or woman has gone before.
- 0.67 We should go into space for the same reason people climb Mt. Everest—because it's there.
- 0.66 Space is the new frontier.
- 0.65 Humans have an innate need to search and discover.
- 0.64 We must broaden our horizons.
- 0.58 Investigation of outer space satisfies human curiosity.
- 0.53 Space offers new challenges, and civilization would stagnate without challenges.
- 0.47 Space exploration fulfills the human need for adventure.

Subfactor 1C: Perspective

- 0.78 Space travel makes us realize that Earth is a fragile, unique, unified world that deserves more respect and better care.

0.74 In space, we see how small our world is and thus learn humility.
0.60 The space program gives us new perspectives on ourselves and our world.
0.54 New experiences and perspectives gained in space inspire art, music, and literature.
0.44 The exploration of space is an unselfish quest that could benefit all mankind.
0.42 Space stimulates the creative, human imagination.

Subfactor 1D: Excitement
0.80 Space travel is fun.
0.77 Space missions are exciting.
0.58 The beauty of space creates a sense of wonder.
0.45 Space gives people something to dream about.

The names of the four subfactors are nothing more than my attempt to capture in a single mnemonic word the theme of the particular group of statements about the value of space exploration. Factor 1 collects the idealistic and emotional justifications for space explorations, thus replicating and expanding greatly upon one result of the Seattle voter study. Some future questionnaire research project should combine items like these with standard personality tests and measures of respondents' propensity to feel various emotions. That is to say, these items seem to reflect fundamental human orientations and feelings, about which individuals differ, that relate to many aspects of life, not just to space exploration or to science and technology.

Note that Subfactor 1A concerns pride, while 1C concerns humility. Perhaps aggressive, achievement-oriented people respond more positively to the 1A Inspiration factor, and passive or introverted people to 1C Perspective. Need for achievement is one of the personality dimensions identified by psychologist David McClelland, and introversion is one end of the extraversion-introversion dimension of personality identified by psychoanalyst Carl Jung long ago and included today in the standard "Big Five" personality dimensions (McClelland 1961; Wiggins 1996). Research on electronic games suggests that some players are by nature explorers, unconcerned with competing against other players, but deriving satisfaction from going beyond the traditional boundaries of their own experience (Bartle 2004). Such people would rate highly Subfactor 1B Exploration. Subfactor 1D Excitement is emotionally positive, but seems less goal-oriented than 1A and 1B.

Frankly, I cannot recall seeing a two-stage exploratory factor analysis that gave results as clear as these. The four subfactors are very different from each other, and the items in each express one primary human values in a coherent manner. Yet the four also fit together, in that Factor 1 is rather more abstract than the other factors we shall consider below, expressing human longings that are not inherently technological and could be satisfied in ways other than through space exploration. In the research concluded in 1972–1986, it seemed reasonable to call these the *emotional-idealistic* values the space program might serve. Here we might call them *psychological* or *motivational*, but I shall use the term *abstractions* to identify the factor, because both the entire group and the four subfactors have abstracted deep meanings from the technological superficialities of rockets and satellites.

4.4 Other Worlds

The second factor that emerged during analysis of the S1986B data brought together statements about human colonization of outer space and exploitation of its resources. This replicates the Colonization factor from the earlier research, and that is an appropriate name for these thirteen statements:

Factor 2: Colonization

0.80 The Earth is too small for us, so we must expand off this planet.
0.79 Space offers room for the expansion of the human species.
0.75 Space settlements could ease the growing problem of overpopulation.
0.75 We could find new worlds we can live on or transform a planet to make it habitable.
0.66 We need an alternate home planet in case the Earth is destroyed by a natural catastrophe or nuclear war.
0.65 Humans should spread life to other planets.
0.63 We could establish manned space stations, communities in space, and space cities.
0.61 Our future ultimately lies in space.
0.58 Eventually, interstellar travel could be possible, taking people to distant stars.
0.51 In space, we could create new cultures, lifestyles, and forms of society.
0.51 We could use raw materials from the moon and planets when natural resources are depleted on Earth.
0.45 We could find new mineral resources on the Moon, Mars, or the asteroids.
0.44 Farms in space and advances in terrestrial agriculture aided by the space program could increase our food supply.

Clearly this group contains a good deal of variation, with the first three most highly loaded items seeking to escape the Earth, and the three at the bottom wishing to bring outer-space resources home to Earth. From the perspective of the average citizen, the Colonization items may seem just as radical as those in the Abstractions Factor, but are more specific in terms of the technical steps that must be taken. While a couple of them seem idealistic, taken as a group they do not seem to express any particular personality type, but rather a general if lofty goal the space program might seek to achieve.

Another way to distinguish the first two factors is to consider the criticisms that could be raised by someone who was dismissive of them. Factor 1 might seem to be foolish, empty rhetoric, expressing moods and feelings but not offering anything capable of technical evaluation. In contrast, Factor 2 could be criticized for being impractical on technical grounds, either because interplanetary travel has proven prohibitively expensive, or because the planets of our solar system have revealed themselves to be hostile for human life and lacking any resources we could not more easily obtain from Earth's own oceans or that Mars-like planet to the south of us, named Antarctica.

Indeed, the feasibility of interplanetary colonization is one of the greatest questions facing humanity. The task seems more difficult today than we thought it might be many decades ago, yet we cannot be sure that broadly-based technological progress might not bring it within the realm of feasibility in some future century.

Prior to the first probes to reach the red planet in the 1960s, science fiction writers imagined that both Mars and Venus might be habitable. The classic image of Mars was an old, desert world that had once been more verdant but might still harbor life, as depicted in *The War of the Worlds* by H. G. Wells in 1898 or *A Princess of Mars* by Edgar Rice Burroughs in 1912 (Wells 1898; Burroughs 1917). Venus, given its heavy cloud cover, was imaged to be a swamp world, teaming with life as in *Pirates of Venus* by Burroughs in 1934 or *Between Planets* by Robert A. Heinlein in 1951 (Burroughs 1934; Heinlein 1951). The Wikipedia page for this latter novel correctly notes: "Like many science fiction works of its period, the novel depicts both Venus and Mars as suitable for human habitation. Since no interplanetary space probes had been launched at the time, neither the extreme pressure and temperature at the surface of Venus, nor the extremely low atmospheric pressure at the surface of Mars, were known to science. Even the length of the day on Venus was not yet known".

In the case of Mars, professional astronomers had long debated how suitable for life the environment might be, with Percival Lowell early in the twentieth century taking the position that both the atmosphere and water supplies were sufficient for life, and some of the seasonal changes dimly seem through telescopes supported the theory that living creatures might be abundant there (Lowell 1906a, b; Lane 2005) In the middle of the twentieth century, many astronomers thought the Martian atmospheric pressure was much lower than that of Earth, but might be just barely high enough to support life, a hope that proved to be very overoptimistic (De Vaucouleurs 1950). The early space probes proved that Mars was not quite as inhospitable as the Moon, but with an atmosphere far too thin to support human life or any of the other living things that thrive on Earth (Horowitz 1986; Carr 1981).

This does not say that colonization is absolutely impossible, but sets very severe constraints that may prevent it from being worthwhile. Give the investment of vast resources, one could conceivably establish subterranean cities, protected from the near-vacuum, intense cold, and solar radiation, laboriously mining resources through the use of machines that could tolerate the environment. Visionaries have occasionally suggested one or another plan for terraforming Mars, to make it livable, but none of the technologies that might conceivably achieve this are sufficiently well developed even to assess their practicality (Allaby and Lovelock 1984).

4.5 The Major Triad

Factors 3, 4 and 5 are much less radical than the first two, and far more grounded in today's science and technology. We shall soon see that they are more popular, and they have much in common with respect to meaning, as should be apparent from reading the items in them:

Factor 3: Technology

0.68 The space program contributes much to our technology.
0.66 Technological spin-offs (advancements developed for the space program, then applied to other fields) improve every-day life.
0.61 The space program contributes to the advancement of science.
0.61 Space research provides valuable, practical information.
0.58 The space program produces better computers, calculators, and electronics.
0.57 The long-term, ultimate benefits of the space program could eventually be important.
0.55 The space program has great benefits for industry.
0.54 Space research tests our scientific theories and promises conceptual breakthroughs.
0.52 Space has great commercial applications and many opportunities for business.
0.48 The space program stimulates the economy and has direct economic benefits.
0.45 Space could offer many unexpected benefits we cannot now foresee.
0.41 In the weightlessness and vacuum of space, we could manufacture new and better alloys, crystals, chemicals, and machine parts.

Factor 4: Information

0.72 Satellites are an important component in navigation systems.
0.71 Satellites are useful in surveying and mapping the Earth.
0.68 Satellites link all corners of the globe in a complete information and communication network.
0.67 Meteorology satellites are great aids for predicting the weather and understanding atmospheric patterns.
0.59 Satellite photography of the Earth contributes to geology, oceanography, and archaeology.
0.59 Communication satellites improve television transmissions.
0.55 Observations from orbit help us find new sources of energy and minerals on the Earth.
0.42 An orbiting space telescope could give astronomers a much better view of the stars.
0.40 In the weightlessness and vacuum of space, we could manufacture new and better alloys, crystals, chemicals, and machine parts.

Factor 5: Knowledge

- 0.70 We could discover our own origins, learning about the history of the universe and Earth.
- 0.67 Through the space program we could learn the origin of life.
- 0.62 We could gain greater understanding of the world we live in.
- 0.61 We could gain a better understanding of the universe as a whole and how it functions.
- 0.51 We could gain knowledge about ourselves.
- 0.43 Space probes increase our knowledge of space, planets, comets, and the entire solar system.
- 0.43 Space research benefits physics —in studies of the nature of matter, for example.

The first and third items in the Technology factor merely name technology and science, and the fourth refers to "valuable, practical information." It is the second item that is distinctive: "Technological spin-offs (advancements developed for the space program, then applied to other fields) improve every-day life." Every year, beginning in 1973, NASA has published a report announcing recent examples of valuable spin-offs to industry, and admits its first function is to serve as "a convincing justification for the continued expenditure of NASA funds". Indeed, it is difficult to determine the extent to which the spin-off reports are spurious propaganda versus factual statements, although members of the spaceflight social movement may be quite honestly convinced they are true (Bijlefeld and Burke 2003). Writing as members of that movement, Ordway, Adams, and Sharpe devoted their 1971 book *Dividends from Space* to spin-offs, but admitted it was often difficult to identify the ultimate sources of an innovation, especially when many fields of technology are closely connected (Ordway et al. 1971). This is especially difficult for the years of the Space Race, when an innovation may have been achieved in one of the military programs but seen the public light of day via transfer to the space program (Hooks 1990). The idea of spin-offs is not new, given that diffusion of innovations has been a topic in the sociology of science and technology for a century (Ogburn 1922; Bauer 1969).

In an as-yet unpublished study done at NASA's request, I examined the agency's claims about medical spin-offs and found three kinds of error that the NASA people might not have noticed. First, equivalent versions of an innovation they attributed to the space program may have been developed long before in another context. Second, a spin-off may be reported too early in the technology transfer process to know whether it really will have lasting significance. Third, even when the space program was involved in the development of a valuable innovation, the contributions of non-space partners may have been far more significant than NASA personnel recognized.

One 2000 study of a NASA effort to transfer innovation to small businesses indicated that it had the negative effect of reducing the companies' investments and innovations in other areas, thus a net zero in promoting overall technological progress (Wallsten 2000). A much more recent study of the National

Nanotechnology Imitative found the same thing for that very different technology being promoted by the federal government (Jung and Lee 2014). While the findings of these studies can be debated, they make the valid point that public investment involves a trade-off, even before it produces a spin-off. Arguably many other fields of pure science and innovative engineering could produce beneficial spin-offs, and one would think that some of them would be far better investments today than space technology, if only because far less has already been invested in them.

Many significant spin-offs cannot be expected from the current NASA manned spaceflight effort, because new technology development is not central to it. Indeed, NASA has cut back research in nanotechnology and robotics to pay for the design and prototyping of launch vehicles that will be "new" in a sense but based on principles developed for the Apollo program over four decades ago. Through the 1970s, development of space and missile technology helped drive the development of computers, but that period is now over and computer science progress is stimulated by developments in a myriad of other areas, from bioinformatics to nanoelectronics, and from home information technology to computer vision for cars. Thus, quite apart from its deeply dubious nature, the Technology factor may be obsolete, until some really new phase in space exploration ignites a fresh wave of innovation.

The Information factor is obsolete in a somewhat different sense. Its claims are almost certainly true, but now refer to past accomplishments of the space program which should be honored but have largely been completed. The Global Positioning System is one of the breakthrough applications of recent decades, based essentially on communication satellites that transmit time signals, and indeed comsats have long played an important role in radio and TV systems. Again, however, the technology is rather mature, and any further improvements are likely to be in aspects that are incidental to space technology. Earth resources satellites have proven to be of moderate value, and undoubtedly they will continue to be launched over the years. However, this application does not require any further development of space technology; sensor technology may improve, but we already know perfectly well how to get satellites into orbit, how to control them, and how to power their onboard equipment. The Hubble Space Telescope launched four years after the Harvard questionnaire data were collected, and certainly has proven that an "orbiting space telescope could give astronomers a much better view of the stars."

The Technology and Information factors share an item, at the bottom of each list, concerning possible advantages of manufacturing in the weightless environment of space. While this idea retains a certain plausibility, to my mind the advantage of weightlessness would be the ability to control better the assembly of small components, even at the microscopic scale. A decade after the Harvard data were collected, a nanotechnology social movement was arising, and today promises far more convincingly to be able to facilitate many more kinds of manufacture than weightlessness could do. Indeed, the key individual in the promotion of nanotechnology before the US government took the lead was Eric Drexler, a disappointed member of the spaceflight social movement who was seeking a new province for his visionary dreams (Bainbridge 2007).

The statements in the Knowledge factor seem convincing, and unlikely to become obsolete in the near future, because we still understand so little about the universe and our place in it. However, there is again the issue of the trade-off in investing the astronomical sums required by space exploration, versus investing in other areas of science. If we really wanted to discover "our own origins," we would be investing heavily in prehistoric archaeology and physical anthropology. The cost of one shuttle launch would have been enough to fund the current effort in these areas for decades.

4.6 Additional Factors

Exploratory factor analysis of large numbers of questionnaire items often delivers many factors of fewer and fewer items, and this analysis generated fully 14. However, each of them seems meaningful. Factors 6 and 7 are especially significant and fundamentally express opposite ideas, the roles space exploration plays in cooperation versus conflict:

Factor 6: Cooperation

- 0.61 The space program generates national unity, encouraging cooperation between numerous sectors of society.
- 0.60 The common cause of space exploration unites the peoples of the world and could eventually create a world community.
- 0.59 The space program contributes to world peace.
- 0.58 Joint space projects between nations improve international cooperation.
- 0.56 Competition in space is a constructive outlet for nationalistic rivalries that otherwise would take the form of aggression and conflict.

Factor 7: Military

- 0.85 A space-based anti-missile, system, part of the Strategic Defense Initiative, could reduce the danger of war and nuclear annihilation.
- 0.84 There are great military applications of space.
- 0.84 The space program contributes to our defense.
- 0.57 Reconnaissance satellites help prevent war and nuclear attack.
- 0.42 The space program builds national pride.

International cooperation in space seems to have been a positive force helping the US and Russia escape the Cold War, from the Apollo-Soyuz Test Project of 1975 in which spacecraft launched by the two nations rendezvoused in orbit, to the current situation in which Russian vehicles provide the only means for human travel to and from the International Space Station. The Strategic Defense Initiative was a US effort launched in the Reagan administration to nullify the danger of Russian ICBMs, and is sometimes cited as one of the factors that led to the disintegration of the Soviet Union. There remains much room to debate the impact of SDI, but Chap. 5 argues the negative on both propositions. Similarly, the idea that joint space missions promote peace is open to doubt, and in the case of US-China relations the interaction that really mattered for improved relations was obviously economic trade. However, the item about reconnaissance satellites had a considerable degree of plausibility back in 1986, and they remain important today (Berkowitz 2011).

The remaining seven factors express smaller ideas, several of which are offshoots of previous factors, most of them suggesting ways in which developments in space could benefit ordinary life here on Earth:

Factor 8: Weightlessness

- 0.75 New medicines could be manufactured in the zero gravity and vacuum of space.
- 0.74 Some medical problems could be treated more effectively in the weightlessness of space.
- 0.72 Medical research performed in space could benefit human health.
- 0.47 In the weightlessness and vacuum of space, we could manufacture new and better alloys, crystals, chemicals, and machine parts.

Factor 9: Resources

- 0.67 New fuels found in space or the development of fusion power in space could help solve the Earth's energy problem.
- 0.51 Solar power stations in orbit could provide clean, limitless energy to the Earth.
- 0.49 We could find new mineral resources on the Moon, Mars, or the asteroids.

Factor 10: Employment

- 0.65 The space program provides jobs for thousands of people.
- 0.59 The space program employs many engineers and scientists who otherwise would not be able to utilize their talents.

Factor 11: Education

- 0.56 Space travel makes us realize that Earth is a fragile, unique, unified world that deserves more respect and better care.
- 0.53 The space program is an educational tool, helping us learn from each other.

Factor 12: Pollution

- 0.80 We could preserve Earth's environment by moving the most polluting industries into space.
- 0.69 The Moon or the sun could be used for safe disposal of toxic materials and nuclear wastes.

Factor 13: Fun

- 0.70 Space travel is fun.
- 0.51 Space missions are exciting.

Factor 14: Probes

- 0.48 Space probes increase our knowledge of space, planets, comets, and the entire solar system.

The last two of these, Fun and Probes, are clearly subfactors. The two Fun items also belong to the Excitement subfactor of Factor 1, so they could even be described as a sub-subfactor. The two others in the Excitement subfactor concern dreams and "a sense of wonder," which is a phrase taken from the standard literary criticism of science fiction. Fun and excitement represent emotional stimulation, while dreams and wonder are more reflective or even passive. The fact that the space probes item appears as its own factor, as well as belonging to the Knowledge factor, may reflect the fact that unmanned probes are a clear alternative to the manned space program, and frequently appear in the news as a separate topic.

Two of the 90 items failed to wind up in any factor, which at first might seem perplexing because each looks like it might fit into one of the factors quite well. One appears to be almost the definition of the Colonization factor: "We could colonize the Moon, Mars, and other satellites or planets of our solar system." What distinguishes this item from those in the Colonization factor is its grandiose quality, and lack of technical specifics. The other unfactored item might fit the tiny Pollution factor: "From space, we could find new ways to control pollution and clean up our environment." But its wording suggests monitoring the Earth's environment, or otherwise intervening in the terrestrial debate about environmental protection, while the two items in the tiny Pollution factor both involve moving pollution from Earth to space.

4.7 Influences on Support for Space Funding

Because questionnaire S1986B contained the space funding item from the General Social Survey, we can examine how the statements about the value of the space program relate to enthusiasm for it. We should be clear at the outset of the following analysis that there are two ways to conceptualize this relationship. First, statements that correlate with support for space development may constitute the causes of that support. For example, a person who is especially worried about the Cold War between the US and USSR might support the Strategic Defense Initiative without having any special enthusiasm for spaceflight more generally. Second, statements may be rationales or reflections for enthusiasm the person already has, perhaps for entirely idiosyncratic reasons. For example, someone who is already a supporter may agree that the program has inspirational value. Both such relationships may have a situational component, such that in the absence of the Cold War SDI may be irrelevant, and a different set of childhood experiences might cause someone to find inspiration in music or humanitarianism, rather than space technology.

A questionnaire is seldom the right methodology to distinguish cause from effect, especially if it is not part of a panel study that administers the same items repeatedly over the span of years to chart processes of change in which cause comes before effect. Yet it can be valuable to identify connections. Indeed, often two variables are connected as mutually reinforcing influences, each both the cause and effect of the other.

In responding to the GSS space funding item, 370 Harvard students felt funding was too little, 345 considered it about right, and 88 called it too much. Another 91 replied "don't know" and will be excluded from this analysis of that item, although their ratings of the 125 items tended to be similar to those from students who felt funding was about right. Table 4.4 lists the 14 factors and the 4 subfactors, plus at the bottom comparing the 90 items that were included in the factor analysis, and the 35 that were not. The "high rating" is the percentage of all respondents who circled a 5 or 6 on the 0 to 6 rating scale, considering the item to be a good reason for supporting the space program. Across all 125 items, 23.8 % of the ratings were 5 or 6, and the mean rating was 2.82, just shy of 3.00 which was the mid-point of the scale.

The first thing worth noting in Table 4.4 is that the items in just 6 of the factors had mean ratings above 3.00: Probes (4.31), Weightlessness (4.19), Technology (4.18), Resources (3.95), Information (3.90), and Knowledge (3.48). Three of these were small factors with few items, and the three substantial popular factors were what above I called the Major Triad. All of these items have a degree of technical clarity and are neither emotional nor visionary. Several of the items, including the one-item Probes factor, had already seen significant accomplishments by 1986.

At the opposite extreme, Excitement (1.91) and its effective subfactor Fun (1.72) have the lowest mean ratings, probably because they are trivial benefits that cannot justify the great investments required by space exploration. The lowest other factor is Military (2.16), which is politically controversial and likely to get less support in the liberal Harvard environment.

In the very bottom row of the table we see that the mean rating given all 125 statements is 3.22 among those respondents who feel too little is being spent on the space program, compared with 2.73 among those who feel funding is about right, and only 1.76 among those who say too much is being spent. Clearly, whatever cause-and-effect relationship we theorize, the justifications for the space program do correlate strongly with support.

The final columns of the table give two alternative simple ways of summarizing this difference, dividing the mean for "too little" respondents by the mean for "too much" respondents, or subtracting "too much" from "too little." High too-little/too-much ratios identify the factors that have low mean ratings. Frankly, such comparisons are somewhat affected by the constraints imposed by the ends of the rating scale. However, we can hypothesize that they also reflect two real phenomena: First, people who are already enthusiastic about spaceflight are especially willing to give favorable ratings to arguments that would not convince other people. Second, justifications that are realistic tend to get good ratings from all respondents, only moderately affected by differences in the degree of personal interest they have in the program.

Table 4.4 Support for space program funding and the meanings of spaceflight

Groups of space exploration goals	Items	High rating (%)	Mean rating by space funding view				Ratio: little/much	Difference: little-much
			All views	Too little	About right	Too much		
Factors								
1: Abstraction	29	16.7	2.43	2.93	2.24	1.25	2.34	1.68
2: Colonization	13	20.5	2.68	3.14	2.57	1.52	2.07	1.62
3: Technology	12	39.4	4.18	4.30	3.70	2.50	1.72	1.80
4: Information	9	39.3	3.90	4.18	3.92	2.93	1.43	1.25
5: Knowledge	7	32.3	3.48	3.85	3.34	2.47	1.56	1.38
6: Cooperation	5	18.1	2.60	3.00	2.52	1.42	2.11	1.58
7: Military	5	15.1	2.16	2.28	2.38	1.12	2.04	1.16
8: Weightlessness	4	48.3	4.19	4.51	4.14	3.21	1.40	1.30
9: Resources	3	42.8	3.95	4.12	3.99	3.07	1.34	1.05
10: Employment	2	16.5	2.64	2.99	2.63	1.60	1.87	1.39
11: Education	2	22.5	2.94	3.21	2.82	2.09	1.54	1.12
12: Pollution	2	20.8	2.59	2.91	2.60	1.61	1.81	1.30
13: Fun	2	12.5	1.72	2.24	1.46	0.65	3.45	1.59
14: Probes	1	51.2	4.31	4.75	4.18	3.16	1.50	1.59
Subfactors								
1A: Inspiration	10	15.2	2.46	2.94	2.35	1.21	2.43	1.73
1B: Exploration	9	18.9	2.54	3.12	2.28	1.29	2.42	1.83
1C: Perspective	6	18.3	2.56	2.96	2.42	1.49	1.99	1.47
1D: Excitement	4	13.3	1.91	2.44	1.63	0.88	2.77	1.56
Groups								
Analysis items	90	25.4	2.97	3.39	2.88	1.86	1.82	1.53
Other items	35	19.7	2.43	2.77	2.36	1.51	1.83	1.26
All items	125	23.8	2.82	3.22	2.73	1.76	1.83	1.46

4.8 A View from the Zenith

The primary value of an in-depth analysis of justifications for the space program is to determine the human values it can reasonably claim to meet. Over time, the plausibility of one or another may change. Some projects will be fully achieved, such as radio communications and navigation Earth satellites, which may still improve in quality but for years have been at least satisfactory. Others may prove technically infeasible or become obsolete as a human goal is achieved by other means, as may be the case for anti-ICBM space-based systems, whose antidote is peaceful coexistence. Still others may be such abstract values that the general public reads them into many quite different areas of human activity, from decade to decade.

Such qualifications aside, the 1986 study seems to have done a good job of documenting the conceptual structure of the spaceflight ideology at its zenith. The Harvard student respondents were born around the time of Sputnik, watched the Apollo lunar landings on television, and had unusually high levels of intelligence and education which allowed them to understand many aspects of space exploration. The Challenger disaster undoubtedly focused the attention of many of them, and caused them to ponder what the space program meant for them, personally, in items of their own values. The following chapter will return to opinion polls of the general public, then examine other specialized respondent polls, first to focus on how other events in the history of spaceflight focused attention in ways that emphasized one or another meaning, and then to concentrate on possible future watershed events.

References

Allaby, M., & Lovelock, J. (1984). *The greening of Mars*. New York: St. Martin's Press.

Bainbridge, W. S. (1978) Public support for the space program. *Astronautics and Aeronautics*, June, pp 60–61, 76.

Bainbridge, W. S. (1991). *Goals in space: American values and the future of technology*. Albany, New York: State University of New York Press.

Bainbridge, W. S. (2004). The evolution of semantic systems. In M. C. Roco & C. D. Montemagno (Eds.), *The coevolution of human potential and converging technologies* (pp. 150–177). New York: New York Academy of Sciences.

Bainbridge, W. S. (2007) *Nanoconvergence*. Upper Saddle River, New Jersey: Prentice-Hall.

Bainbridge, W. S. (2014). *Personality capture and emulation*. London: Springer.

Bainbridge, W. S., & Wyckoff, R. (1979). American enthusiasm for spaceflight. *Analog, 99*(7), 59–72.

Bartle, R. A. (2004). *Designing virtual worlds*. Indianapolis, Indiana: New Riders.

Bauer, R. A. (1969). *Second-order consequences: A methodological essay on the impact of technology*. Cambridge, Massachusetts: MIT Press.

Berkowitz, B. (2011). *The National Reconnaissance Office at 50 years: A brief history*. Chantilly, Virginia: Center for the Study of National Reconnaissance.

Bijlefeld, M., & Burke, R. (2003). *It came from outer space: Everyday products and ideas from the space program*. Westport, Connecticut: Greenwood.

Burroughs, E. R. (1917). *A princess of Mars*. Chicago: A. C. McClurg.

Burroughs, E. R. (1934). *Pirates of Venus*. New York: Grosset and Dunlap.

Carr, M. H. (1981). *The surface of Mars*. New Haven, Connecticut: Yale University Press.

De Vaucouleurs, G. (1950). *The planet Mars*. London: Faber and Faber.

Glaser, B. G., & Strauss, A. L. (1967). *The discovery of grounded theory*. Chicago: Aldine.

Heinlein, R. A. (1951). *Between planets*. New York: Scribner.

Hooks, G. (1990). The rise of the Pentagon and U.S. state building: The defense program as industrial policy. *American Journal of Sociology, 96*(2), 358–404.

Horowitz, N. H. (1986) Mars: Myth and reality. *Engineering and Science*, March, 4–9, 35–37.

Jung, H. J., & Lee, J. (2014). The impacts of science and technology policy interventions on university research: Evidence from the U.S. National Nanotechnology Initiative. *Research Policy, 43*, 74–91.

Lane, K. M. D. (2005). Geographers of Mars: Cartographic inscription and exploration narrative in late Victorian representations of the red planet. *Isis, 96*(4), 477–506.

Lowell, P. (1906a). *Mars and its canals*. New York: Macmillan.

Lowell, P. (1906b). *Mars as the abode of life*. New York: Macmillan.

McClelland, D. C. (1961). *The achieving society*. Princeton, New Jersey: Van Nostrand.

Ogburn, W. F. (1922). *Social change with respect to culture and original nature*. New York: B. W. Huebsch.

Ordway, F. I., Adams, C. C., & Sharpe, M. R. (1971). *Dividends from space*. New York: Thomas Y. Crowell.

Wallsten, S. J. (2000). The effects of government-industry R&D programs on private R&D: The case of the small business innovation research program. *The Rand Journal of Economics, 31*(1), 82–100.

Wells, H. G. (1898). *The war of the worlds*. New York: Harper.

White, H. C., Boorman, S. A., & Breiger, R. L. (1976). Social structure from multiple networks. *American Journal of Sociology, 81*, 730–780.

Wiggins, J. S. (Ed.). (1996). *The five-factor model of personality: Theoretical perspectives*. New York: Guilford Press.

Chapter 5
Events

Post-Apollo events both inside and outside the space program have been the focus of numerous opinion polls. In most cases, the public expressed some interest and a degree of encouragement for NASA, and a few cases also reveal general qualities of public spaceflight opinion. Other polls asked about events that had not yet happened, notably questions about possible future manned Mars missions, just as the pollsters had asked about Moon missions prior to the end of the Space Race. Both a return to the Moon, and an expedition to Mars, are goals that are very much in limbo at the present time, as no grand manned mission seems affordable given the financial difficulties faced by government and the absence of compelling motivations to undertake them.

This chapter will set the background for consideration of specific events by charting changes in space funding opinions over the full history of the General Social Survey data, then examine two event-related issues that were salient during the Reagan presidential administration: (1) the Strategic Defense Initiative which was an event triggered from outside the space program by a political shift in US relations toward the Soviet Union, and (2) the Challenger shuttle accident that was endogenous to the space program and required a decisive response. A possible return to the Moon will be considered through input from scientists and engineers collected through a form of questionnaire.

Wernher von Braun and his colleagues outlined technical specifications for a Mars expedition in some detail around the year 1950 (von Braun 1953). However, I do not see much evidence that the recent general public understands very much about how such an endeavor might be accomplished, despite the existence of a small Mars-oriented social movement, so their responses may reflect general values rather than well-grounded opinions (Zubrin 1996). The public has little basis to judge missions involving asteroids, and may feel that Apollo accomplished much of what any lunar expeditions could do, while Mars is at least a powerful concept in the public mind.

W.S. Bainbridge, *The Meaning and Value of Spaceflight*,
Space and Society, DOI 10.1007/978-3-319-07878-6_5

5.1 The Recent Past

The General Social Survey question about funding for the space program, administered every year or two from 1973 through 2012, offers a high-level perspective on changing attitudes, that may help us understand the impact of events during that period. Figure 5.1 graphs the three responses, expressing whether respondents feel funding for the space exploration program was too little, about right, or too much. The most obvious change in the graph is the consolidation of attitudes over the first few years, as the percent saying too much was being spent dropped from 61.4 % in 1973 to 42.7 % in 1980, allowing the other two responses to increase. Those feeling about the right amount was being invested rose from 30.8 to 37.6 %, and those saying too little rose from 7.8 to 19.6 %. After peaking at 19.6 % in 1980, those feeling too little was being invested dropped, hitting other peaks in 1988 at 18.9 %, in 2000 at 15.0 %, and in 2012 at its all-time high of 22.6 %. Perhaps most striking, visually, is something we have long known in terms of the numbers: the fraction wanting space program funding increased has always been far less than the two other opinions.

We might hypothesize that the increase in support for the space program at the beginning of the period might have reflected a more general recovery of enthusiasm for American institutions after the end of the Vietnam War and the fading of the counterculture of the 1960s (Pion and Lipsey 1981). However, other GSS variables do not show the same pattern. One battery of items began: "I am going to name some institutions in this country. As far as the people running these institutions are concerned, would you say you have a great deal of confidence, only some confidence, or

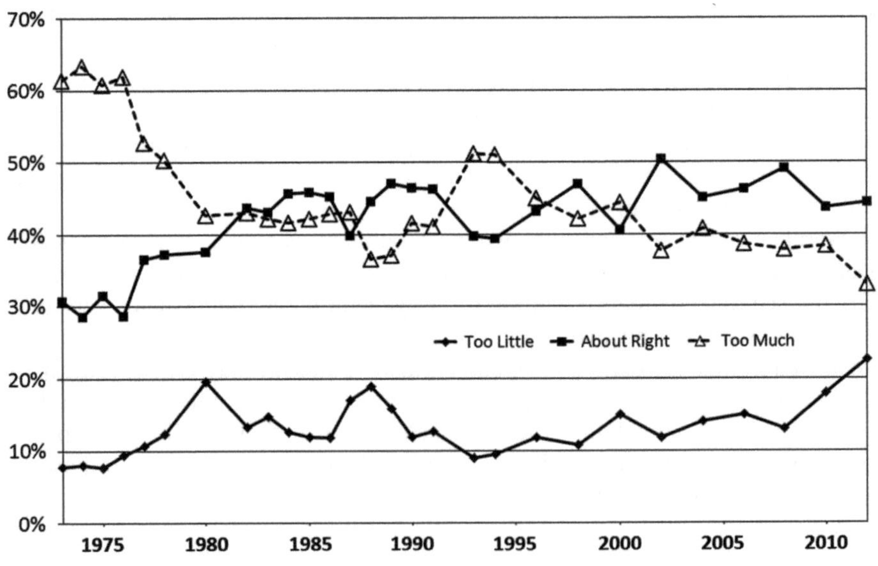

Fig. 5.1 Attitudes toward funding of the space exploration program, 1973–2012

hardly any confidence at all in them?" The percent expressing a "great deal" of confidence in the scientific community was 40.8 % in 1973, and 45.9 % in 1980, but this was not a simple rise but more general volatility, as the number had been 50.4 % in 1974, 42.2 % in 1975, and 48.6 % in 1976. In 2012 it was only 41.9 %. Confidence in the executive branch of the federal government showed a different pattern, 29.9 % having a great deal in 1973, dropping to 13.9 % in 1974 during the Watergate scandal, only at 12.5 % in 1980, and hardly better at 15.0 % in 2012.

In the absence of compelling evidence for some other theory, we can tentatively assume that the consolidation of attitudes about the space program in the 1970s really did represent adjustment to the idea of continuing space exploration efforts after the Apollo Program. Perhaps the 1980 peak was anticipation for the Space Shuttle, which made its first orbital flights in 1981, after which enthusiasm dropped somewhat to a level that would be the norm for the next decades, with 13.3 % feeling too little was invested in 1982. The peak of enthusiasm 1987–1989 came at the end of the Reagan administration, and the following section will consider two events in that period, the Space Shuttle accident in 1986 and the Strategic Defense Initiative. But before we examine specific events, we need to think more deeply about the consolidation at the beginning, leading to a fundamentally stable distribution of attitudes for several years, ending with an apparent upsurge.

Questionnaire respondents react superficially to news events in the surrounding society, but much more deeply to the events of their own personal lives. Chief among them is the point during the life course, perhaps extending from the teens through the twenties, when the patterns of their own lives consolidate, and their general attitudes may become relatively stable. We have already seen that age is a significant factor in shaping spaceflight attitudes, and now we can examine how age and historical events interact. Obviously, age is calculated by subtracting birth year from the year in question. Those born in the same year are said to be members of a *birth cohort*, and the point in time when age or some other variable is measured is the *period* (Davis 2013). Some factors, such as puberty and retirement, are primarily age variables. Period can matter, as for example someone entering adolescence during the Great Depression, and the transition to adulthood during the Second World War, may wind up with very different fundamental assumptions about life than someone who went from adolescence to adulthood in the prosperous but culturally unstable 1960s. We can apply such thinking to the period of the Space Race, although recognizing that social scientists have found it often difficult to disentangle age, period, and cohort effects (Hall et al. 2005).

Table 5.1 separates respondents into categories based on their birth cohort and the year in which they answered the GSS question about funding for the space program. The data for 1989–1991 and 2008–2012 are based on combining three administrations of the GSS, and the other periods are based on two each, in order to have at least 2,000 respondents at each time period. Cells are empty when the total number of respondents in the category was fewer than 100. Looking down the columns, we see the familiar age factor, with the oldest respondents expressing less support for space exploration. Except for the oldest people, we see increases of enthusiasm in each birth cohort in the 1982–1983 data, and in the 2008–2012

Table 5.1 Percent of GSS respondents who say space funding is too little

Birth year of respondent	Years when General Social Survey was administered				
	1973–1974	1982–1983	1989–1991	1998–2000	2008–2012
1890–1909	5.9 %	5.3 %			
1910–1929	7.2 %	9.2 %	5.9 %	4.7 %	4.7 %
1930–1949	8.6 %	15.5 %	14.0 %	12.5 %	16.9 %
1950–1969	10.2 %	17.7 %	17.2 %	14.0 %	18.7 %
1970–1989				15.7 %	18.7 %
Respondents	2,839	3,229	2,040	2,583	2,768

data, which indicates that these two peaks in Fig. 5.1 were not simply the result of changing numbers of respondents in each cohort, but may represent a real if modest boost in public enthusiasm.

Admittedly, it would be better to have data from each year across the period, and larger numbers of respondents in the cells. For example, the 4.7 % for the 1910–1929 birth cohort in 2008–2012 is based on just 107 respondents, compared with 298 in the 1998–2000 period. In a 2009 essay reflecting on the long-term impact of Sputnik, Deborah Stine reported that a public opinion survey had found a powerful cohort effect: Respondents who had come of age during the Space Race were more supportive than younger respondents (Stine 2009). We really do not see that in Table 5.1, but a different subset of the GSS data hint at such an effect. In the 1984 GSS, 19.9 % of the 1940–1959 birth cohort felt too little was being invested, compared with only 9.3 % of the 1960–1979 birth cohort. The pattern was similar but weaker for the same cohorts in the 1994 GSS, 12.2 versus 10.3 %, and in 2004, 15.3 versus 14.2 %. The most recent GSS data suggest that this Sputnik Effect may have faded away, and people of all ages except the elderly are about equally open to appeals for support of the space program.

5.2 The Two Reagan Policies

President Ronald Reagan faced two major decisions related to spaceflight, the first resulting in the Strategic Defense Initiative (SDI) in 1983, and the second was the decision in the wake of the Challenger disaster in 1986 to build an additional Space Shuttle and indeed not to abandon this mode of achieving orbit. While Reagan gave moderate support to the plan to build what became the International Space Station, this project was NASA's own next step in the development of space infrastructure, and thus not an unusual development that fired up public opinion.

The meaning of the SDI was hotly debated in the 1980s, and even today some questions linger. My 1976 book, *The Spaceflight Revolution*, discussed the possibility that the spaceflight social movement could gain ground by promoting a technological solution to the danger posed by nuclear missiles, and back then it was possible at least to imagine if not to design high-reliability defenses against ICBM (Bainbridge 1976). I noted that spaceflight pioneer Eugen Sänger—Wernher von Braun's rival in Germany during the Second World War—had proposed such a system in his 1958 book, *Raumfahrt: Technische Űberwindung des Krieges* (*Spaceflight: Technical Overcoming of War*) (Eugen Sänger 1958).

Sänger's key idea was the old science fiction concept of ray guns, which had appeared in *The War of the Worlds* by H. G. Wells, way back in 1897 (Wells 1898). Today we would conceptualize such a defense in terms of powerful laser beams that could destroy warheads while they were still high in space. The chief alternative has always been firing an anti-missile missile at the warhead, but a technologically advanced enemy could include many decoys, perhaps just a swarm of metal-clad balloons that would appear to radar just like the warheads themselves, at least until they reentered the atmosphere, which would be too late. A laser system could fire repeatedly, destroying decoys as well as warheads. That requires it to be powerful and recharge quickly, which suggests it might be a large ground-based emplacement. However, the atmosphere diffuses laser beams, especially in bad weather, so a laser defense would best be placed in orbit, which requires development of a substantial space launch infrastructure, which then could also be applied to civilian and scientific purposes.

It has always been hard to evaluate the technical issues, because so much of the relevant information is secret. As evidenced by somewhat successful missile defense systems deployed in Israel, ground-based anti-missile missiles appear to work, at least over short ranges against incoming unsophisticated missiles. I doubt that this method has even now been developed for successful defense against long-range, sophisticated weapons, and one occasionally hears of unsuccessful tests which are especially discouraging when one thinks that full national defense against nuclear missiles would need to be highly reliable. However, much of the political science analysis of the Strategic Defense Initiative has viewed it as a move in a propaganda game, or at best a very limited deterrent rather than an effective defense (Slater and Goldfischer 1986; Gromoll 1987; Beth and Fischer 1997). Similarly, while development continues, operational high-power laser weapons apparently have not been developed, over half a century after Sänger wrote, and three decades after Reagan launched SDI.

A rather unusual 1986 journal article by Thomas Graham and Bernard Kramer, "ABM and Star Wars," consists almost entirely of item responses from 64 opinion polls that asked about different conceptions of SDI. The authors had to admit that the general public was not well informed: "Approximately 75 % of the public holds the (inaccurate) belief that the United States has a fairly effective defense against nuclear weapons… In addition, most people are satisfied with our (nonexistent) defense against nuclear attack… In the 1960s, a majority of the population believed that Russia had an existing ABM defense (Graham and Kramer 1986)." ABM stands for Anti-Ballistic Missile, which I like to define as missiles against missiles, while Star Wars is named after the movie series, and I prefer to use this term for the energy beam weapons. Yet in line with government secrecy and public ignorance, perhaps none of the terms really have exact definitions.

These dim facts should not, however, discourage us from looking at SDI public opinion data, because conceptions in this area do influence public perceptions of the meaning of spaceflight more generally. We should, however, keep in mind three points: (1) Ambitious versions of SDI may be technically infeasible, certainly in the 1980s, probably today, and perhaps forever. (2) SDI may have had a role in international relations, during the late years of the Cold War, quite apart from its technical feasibility. (3) The relevance of SDI for spaceflight depends very much on which technologies it would use, and thus what new innovations and infrastructure development would be required that might have secondary value for other space applications.

A fascinating but frankly somewhat confusing set of results can be drawn from a Harris poll taken on December 9, 1987, including two items about SDI. The first asked, "Do you think that SDI, commonly known as the Star Wars plan, when fully developed as a defense against nuclear attack, will really work, or not?" Of 1,250 respondents, 43.1 % said it will work; 46.4 % said it will not, and 10.5 % were not sure. The second asked, "Will SDI be helpful or harmful to the security of the United States?" Fully 61.7 % said it will be helpful, 27.8 % said harmful, and again 10.5 % were not sure. These results seemed to contain a contradiction. Apparently many people felt SDI would fail to work yet nonetheless be helpful.

If we were asked this question retroactively today, we might however say that SDI could not have worked, because even 30 years later effective means for intercepting ICBM warheads have not been developed, yet the Soviet Union collapsed, and the world seems safer. Some may believe that SDI hastened the end of the USSR, by making its leaders falsely believe the US had achieved some secret breakthroughs that would render SDI workable, although the USSR itself did not know how to accomplish their own effective version of SDI, perhaps simply because they did not have enough money to develop one. Or, it may be that many respondents were skeptical yet optimistic, expressing doubts when asked if SDI would work, but expressing hopes when asked if it will be helpful.

A fairly substantial social scientific literature argues that the end of the Cold War and the abandonment of the Soviet Union was not in fact a response to anything the United States did (Zakaria 1990; Collins 1995). A classic social scientific theory called the *Iron Law of Oligarchy*, notably proposed by German sociologist Robert Michels on the eve of the First World War, asserts that all relatively stable societies tend to be dominated by elites, neither democratic nor totalitarian (Michels 1915). That theory would have predicted that both Russia and China would have evolved away from totalitarian Marxism, so long as relations with other societies were relatively stable for a number of years. That is exactly what we have seen, and both are rather obviously dominated by ruling elites. So long as international relations do not deteriorate, the elites in both Russia and China can be expected to support the general world economic order, simply because they personally profit from it.

A sense of the real meaning of the Harris questions, and indeed hints of the complexity of that meaning, are suggested by Table 5.2. The 12 items that define the rows of this table were introduced by this question: "In 1988, do you think these things will happen, or not." Examining the first one will illustrate how to read

Table 5.2 The context of the Strategic Defense Initiative

Opinion	SDI will work if		SDI will be helpful if	
	This happens (%)	Does not (%)	This happens (%)	Does not (%)
President Reagan and Soviet Leader Gorbachev will hold another summit in Moscow and will agree to *major* reductions in long-range nuclear weapons	46.2	36.3	64.1	59.3
The economy will slump, with unemployment rising and prices also rising	39.0	50.5	57.5	69.5
Major steps will be taken toward reducing the federal deficit	49.3	37.9	66.1	58.5
The stock market will continue to decline, but will not have another crash	40.5	49.9	62.6	64.1
The Democrats will continue to control the U.S. Senate and the House of Representatives	43.7	49.8	62.5	66.7
The Republicans will continue to control the White House	52.1	35.5	65.6	60.2
There will be a major earthquake, the worst in 80 years, along the West Coast	45.1	42.9	58.0	65.7
The nuclear arms control agreement with the Soviet Union will be ratified by the Senate	46.5	37.0	65.3	56.1
Confidence in President Reagan will continue to decline	36.0	54.3	55.3	71.4
The divorce rate will continue to decline	45.1	42.2	63.5	60.8
The value of the dollar will continue to fall abroad	42.0	48.2	58.9	68.4
The U.S. will end up in a war against Iran	39.9	45.3	54.6	64.9

each row. Discounting one respondent who failed to answer, fully 65.5 % believed Reagan would have a successful summit meeting with USSR president Mikhail Gorbachev, reducing the two nations' stockpiles of long-range nuclear weapons, a goal parallel to that of SDI itself. Another 28.9 % predicted that a successful summit would not take place, and 5.6 % said they were not sure. Among those who predicted a successful summit, 46.2 % felt that SDI would work, while those who made the opposite prediction were nearly 10 % points lower in believing that SDI would work, 36.3 %. There was a difference of the same kind, if somewhat smaller, on the question of whether SDI might be helpful, 64.1 % versus 59.3 %.

Looking down the columns of figures for whether SDI will work, the two predictions showing the biggest differences also relate to Reagan, concerning whether Republicans will continue to hold the White House (after the 1990 election) and whether Reagan's popularity will decline. So one pattern we see in the table is that support for Reagan is associated with confidence in the SDI program that Reagan proposed. Indeed, the item about confidence in Reagan may really express not the abstraction, public confidence, but the personal confidence of the respondent. Of those who reject a decline in public confidence, thereby expressing their own confidence, 71.4 % believe SDI will be helpful. This may not be merely about political biases. Even well-educated and science-attentive respondents may not have possessed the information required to make an independent assessment of SDI, and so their opinion depended upon whether they trusted Reagan (Mondak 1993).

Two items showing little variation actually suggest an interesting finding, the ones about a major earthquake (bad) and a decline in the divorce rate (good). In both cases, agreeing a bad event or a good event will happen correlates with a slightly higher fraction believing SDI will work. This may simply reflect a small degree of *yea-saying bias*, the tendency to agree with whatever the interviewer says (Couch and Kenniston 1960). The pattern is different for the question about whether SDI will be helpful or harmful. Looking though all the predictions of objectively harmful events, one sees a bigger than average difference in favor of SDI being helpful, perhaps identifying a subset of the respondents as innate optimists, rejecting the likelihood of bad events, and more ready than other respondents to believe SDI—or anything else—might be helpful. Some analysts have suggested that the pattern of responses to questions about SDI often represented primarily the extent to which the respondent palpably feared nuclear war (Smith 1988; Russett 1990–1991; Bartels 1991).

While SDI was closely associated with the Reagan administration's own policies, the issue of how to respond to the Challenger disaster on January 28, 1986, would have faced any president. As mentioned earlier, I was at Jet Propulsion Laboratory on that day, and immediately began a sociological research study in response. It was interesting in the following months to observe the investigations that led to criticism of NASA and even sociological analysis of how leaders failed to manage the Shuttle program properly (Vaughan 1990; Heimann 1993). Recalling the chaotic and politically-shaped process through which the Shuttle was designed, however, I came to the view that management was a secondary issue, because the Shuttle design was fatally flawed, and both Shuttle disasters might not have happened had the geometry of the vehicle been different. Specifically, placing the orbiter beside the solid boosters and external fuel tank put it in a vulnerable position. Of course, there were reasons for this, starting with the cost-based decision not to invest in developing a fully independent and reusable booster stage, which could have taken any of several forms.

Thus, the chief decision in the wake of the Challenger disaster was a difficult one: Continuing the program as it was but with better management, versus deciding the design was fundamentally flawed and phasing it out, as eventually was done, in the context of too-little funding to develop a better successor. A more sophisticated analysis might suggest that management was following the culturally dominant paradigm of assuming that the technical problems of manned spaceflight were less than they in fact were, and strict technical realism was incompatible with any manned space program.

The peak in public support for space exploration funding in 1988, quite visible in Fig. 5.1, probably reflects agreement that the Shuttle Program should continue, with the cost of an additional orbiter. Back on September 4, 1986, Harris had included two sets of questions in a poll of 1,255 Americans, and both the questions and responses suggest the meaning of the space program at that time. Three questions had this introduction: "Earlier this year, the space shuttle Challenger blew up just after it took off. This was followed by an extensive investigation and report by the specially appointed Rogers Commission to determine what went wrong and what should be done about the space program." Here are the percentages feeling that each statement was true:

72.8 % The full story of what happened in the disaster of the Challenger has come out
26.0 % NASA has made, or is making, the basic changes necessary to get the space program back on track
20.2 % Those to blame for the Challenger disaster have been identified and have been fired or their companies fired

At the risk of overstatement, one could suggest that the correct answer to the third question is *true*, because the person responsible was President Richard Nixon who was not prepared to invest enough for a better design of the system in the first place, and he was forced out of office—admittedly for other misbehaviors. However, the framing of the question in terms of blame for individuals ignores the possibility that no available technological system could provide really safe human spaceflight at an affordable cost at that time in the evolution of human society.

A set of five additional questions had this introduction: "After considering the Rogers Report and the future of the space program, President Reagan finally decided that he wanted to build another space shuttle vehicle at a cost of $2.8 billion to replace the Challenger that blew up. However, the new space shuttle would not fly until 1992. At the same time, the President does not think that, to pay for the new shuttle, spending on the space program will have to be increased. Let me ask you your reaction to the new Reagan space policies." Here are the percentages agreeing with each statement:

20.2 % It will be possible to pay for the new space shuttle out of the current NASA budget and that extra money to pay for it won't be needed
48.4 % The president was right to ban the shuttle from getting paid to launch commercial satellites because doing that could prove to be too hazardous

39.2 % By making a new shuttle the major expenditure for the space program, it will mean that such important other programs as expendable rockets, reusable unmanned vehicles, and space stations will now be neglected

72.3 % The main emphasis in the space program now likely will be the military, especially the Air Force, which will have the money to finance shuttle trips and to undertake programs that will move toward new discoveries in space

47.8 % Because NASA will be trying to make up for the time lost due to the Challenger disaster, it might well find itself involved in space experiments that look more like the past than the future

In response to another question, 67.6 % of respondents felt the space program is worth the investments made in it, and another question asked: "Some critics of the White House policy on the space program believe that with the set-back from the Challenger disaster, it is critical for the U.S. to spend more money on space, even doubling the budget for NASA, beginning next year. Do you agree or disagree with these critics?" While 4.0 % of respondents were not sure, 36.8 % agreed, while 59.2 % disagreed. Although the majority were not in favor of greatly increased investment, the faction of respondents who were is rather high, suggesting that indeed the modest jump in support for increased funding found in the GSS really did represent a movement among some Americans to respond to the Challenger disaster with increased efforts.

5.3 The Return to the Moon

On December 4, 2006, the National Aeronautics and Space Administration issued several pieces of information about its rationale for returning to the Moon and setting up a base, perhaps at the south lunar pole. In preparation, NASA solicited input from 1,000 scientists, engineers, entrepreneurs, space advocates, and members of the general public, and from a dozen other space agencies. The resulting plan has six themes, listed in Table 5.3, which might be conceptualized as general values that a lunar base might serve, or broadly-defined benefits it might provide. Note that this analysis was based on something very much like a public opinion survey, but one directed at space scientists and engineers, rather than ordinary citizens.

Crosscutting these six themes were 181 objectives, each of which may be connected to more than one theme, and the average is 2.57 themes. The individual objectives are collected into 23 categories: Astronomy and Astrophysics (9 objectives), Heliophysics (8), Earth Observation (12), Geology (16), Materials Science (3), Human Health (8), Environmental Hazard Mitigation (17), Operational Environmental Monitoring (3), Life Support and Habitat (9), General Infrastructure (4), Operations, Testing and Verification (8), Power (3), Communication (5), Position, Navigation and Timing (6), Transportation (4), Surface Mobility (3), Crew Activity Support (5), Lunar Resource Utilization (11), Historic Preservation

Table 5.3 The six themes of NASA's Moon base project

Theme	Definition	Objectives
Human civilization	Extend human presence to the Moon to enable eventual settlement	114
Economic expansion	Expand Earth's economic sphere, and conduct lunar activities with benefits to life on the home planet	96
Scientific knowledge	Pursue scientific activities that address fundamental questions about the history of Earth, the solar system and the universe—and about our place in them	87
Exploration preparation	Test technologies, systems, flight operations and exploration techniques to reduce the risks and increase the productivity of future missions to Mars and beyond	85
Public engagement	Use a vibrant space exploration program to engage the public, encourage students and help develop the high-tech workforce that will be required to address the challenges of tomorrow	30
Global Partnerships	Provide a challenging, shared and peaceful activity that unites nations in pursuit of common objectives	10

(3), Development of Lunar Commerce (9), Commercial Opportunities (14), Global Partnership (4), Public Engagement and Inspiration (17).

For example, the Human Civilization theme sees the lunar base as the first step toward colonization of the moon. One of the 114 objectives that supports accomplishment of this theme is #43, "mGEO14" in the Geology category. In a 23,000-word spreadsheet made available on the web, NASA defined each objective ("name"), summarized the goal, and offered a brief justification saying why it would be worth accomplishing. Here, for example, is mGEO14:

Name: Characterize potential resources to understand their potential for lunar resource utilization

Summary: Locate and quantify (develop maps at appropriate scales) surface/near-surface deposits of potentially valuable resources, including both minerals and volatiles (especially water)

Value: Future exploitation of lunar resources is facilitated if a global surface map of these resources exists. Such a resource map is also of scientific value in helping to define variations in surface compositions

The 181 objectives are not official NASA goals, but rather cogent ideas that had been submitted to the fact-finding process and would be worth considering as the program developed. Hopefully, some foreign governments, corporations, or even non-profit organizations might wish to adopt one of these objectives and work cooperatively with NASA to achieve it. Several are challenges that plans for a lunar base would need to address. In a way, the list is an inventory of what early twenty-first century spaceflight culture thinks about lunar exploration and exploitation, and thus the NASA spreadsheet represents a kind of social science research result. Each of the six major themes can be understood in terms of the objectives associated with it, and a very preliminary analysis follows.

It is difficult to distinguish two of the themes, Human Civilization from Economic Expansion, without an analysis of the objectives connected with them. Sixty of the objectives serve both themes, while 36 serve Economic Expansion but not Human Civilization, and 54 serve Human Civilization but not Economic Expansion. A key distinction concerns whether expanding civilization to the moon and planets is a good thing in itself, or must return clear profits to the Earth.

The distinctive *Human Civilization* objectives include many that concern technical solutions for living in the exotic lunar environment, including handing radiation bombardment from the sun and galaxy, enduring wide temperature extremes, defending against micrometeorites, dealing with surface dust that may insinuate itself into equipment and living spaces, and managing the lack of air and concomitant need for pressurized suits, vehicles, and habitats. Some objectives directly cite investments that must be made: "Emplace support services on the Moon, including emergency response, to enable increased lunar activities." "Implement a secure, reliable, robust, interoperable and scalable telecommunications capability to support expanding telecom needs of exploration operations." Others envision using lunar resources to avoid having to ship everything at great cost from Earth: "Develop and validate tools, technologies, and systems that excavate lunar material, to enable lunar resource utilization." "Produce propellants and life support and other consumables from lunar resources, to improve the productivity of lunar operations." Assuming eventually millions of people will live on the Moon, environmental protection policies should be established from the very beginning: "Preserve regions of the Moon in their natural state to protect them from developing lunar activities."

The term *Economic Expansion* may convey too narrow a concept, because many of the objectives serving this theme concern human welfare on Earth, broadly defined, rather than just commercial profits. Some objectives concern moon-based activities to understand external threats to the Earth: "Detect and monitor Near Earth Objects (NEO) to discover threats to the Earth and Moon." "Analyze the Sun's role in climate change to gain a better overall understanding of climate." Others monitor the changing conditions on our planet: "Observe the Earth's atmospheric composition to characterize its dynamics." "Measure the Earth's ocean color to understand its health." Strictly economic objectives are also covered by this theme: "Identify and enable commercial markets, based on lunar activities, to broaden the scope and value of lunar activities." "Utilize innovative commercial entertainment and media outlets to broadcast to the public high-bandwidth video, imagery, and other information, to generate revenue and engage the public." "Develop interactive video games based on lunar exploration to generate revenue and engage the public." "Emplace items on the Moon that can be controlled remotely by the public on Earth to generate revenue." Among the Commercial Opportunities imagined are earning money from rich tourists, selling new materials to Earth, and possibly transmitting electric power to Earth.

Many of the *Scientific Knowledge* objectives envision that the moon will become an observational platform for studies of the sun, interplanetary space, the geology or natural environment of the moon itself ("selenology"), materials science, biology (human health and astrobiology), robotics, and Earth observation. The social and behavioral sciences are not mentioned prominently, although there are hints about studying human cognitive performance under lunar conditions and "understanding the impact of isolation on crews." One practically-oriented objective says, "A systematic, comprehensive set of operations testing will characterize how fundamental living and working tasks are best accomplished in the lunar environment." Another says, "Human factors research aims to understand whole body coordination strategies, including balance, posture, locomotion, work capability, endurance, and speed of humans in fractional gravity, and the effect of isolation and communication lag on performance and mission coordination."

The *Exploration Preparation* theme was fundamental to the current redesign of the American space program, intending to go back to the Moon as a step toward Mars. The spreadsheet uses the word "Mars" 67 times in 40 different objectives. Martian gravity, the Martian atmosphere, and Martian terrain fall between the extremes on the Moon and Earth. To a first approximation: If we can handle the Moon, we can handle Mars. Two related objectives posit the revolutionary possibility of learning how to live without the nurturing biosphere of our home planet: "Closed loop life support systems enable long duration human settlement and exploration missions, including Mars missions, by providing the capability for self-sufficient operations with minimal impact on the surrounding environment." "Successful closure of the life support system with ecologically balanced plant and microbial communities will reduce resupply logistics for extended lunar stays and Mars missions and facilitate significant terrestrial benefits, particularly in waste management." Mars is not the only goal: Lunar materials could possibly be used in cis-lunar space activities, and the fuel for space exploration vehicles might come from the Moon.

Eleven of the thirty *Public Engagement* objectives are not also coded under one of the other themes, and thus give a clear picture. One states the general idea: "Provide education, communication, engagement, and outreach activities to assist in building general knowledge of, and ultimately support for, lunar exploration activities." Among the specific methods considered are opportunities for Earth-bound people to communicate with people on the Moon, projects allowing students to contribute to lunar activities, and porting space-related video through new Internet-based distribution sites. One objective apparently recognizes that the spaceflight social movement was a significant source of support in the past and could be again: "Utilize existing pro-space and student organizations to galvanize public support about exploration."

The *Global Partnership* theme involves just 10 objectives, most of which concern international coordination of efforts: "Establish a global partnership framework to enable all interested parties (including non-space faring nations and private companies) to participate in lunar exploration." "Establish standards and common interface designs to enable interoperability of systems developed by a global community." "Establish a set of export control laws and regulations that will enhance effective global cooperation on lunar activities." "Explore new methods of collaboration between and among industry, government, and academic entities, to maximize the benefits that each can bring to each other." "As necessary, establish appropriate legal mechanisms for lunar surface and orbital activities to enable commercial and governmental involvement." Presumably, cooperation in these rather bureaucratically technical areas could cement relations between the nations and organizations, thereby contributing indirectly to world unification. As a symbol of shared human aspirations, a lunar base could contribute directly to a sense that we are one species, joining together to move out into the wide universe.

5.4 Mars Missions

In the years after NASA published the results of its remarkably detailed Moon study, a new administration in Washington abandoned any plans to return there, without substituting any other clear objective. Of course the longstanding assumption of the spaceflight social movement was that a Mars expedition should follow the conclusion of lunar exploration. In 2012, the National Research Council observed: "A human mission to Mars has been the ultimate goal of the U.S. human spaceflight program. This goal has been studied extensively by NASA and received rhetorical support from numerous U.S. presidents, and has been echoed by some international space officials, but it has never received sufficient funding to advance beyond the rhetoric stage. Such a mission would be very expensive and hazardous, which are the primary reasons that such a goal has not been actively pursued (National Research Council 2012)."

Several diverse questionnaire datasets exist that can be used to evaluate public conceptions of the exploration of the red planet. The 1998 Southern Focus Poll, using weighting procedures to combine respondents both inside and outside southern US states, reports that 74.6 % of respondents selected "favor" in response to this item: "And now a question about space travel. Would you favor or oppose sending a manned rocket to land on Mars?" Support was somewhat higher among men than women, 77.1 % versus 72.6 %. There also were small political differences, with 78.4 % of Republicans favoring, compared with an almost identical 78.0 % of Independents, and a slightly lower 73.7 % of Democrats.

As we have seen with other space variables, there was a substantial difference by levels of education in the Southern Focus Poll. Just 51.4 % of those with less than high school educations were in favor, compared with 66.1 % of high school graduates, 75.8 % of those with some college, and 86.5 % of college graduates. The percent in favor dropped slightly to 84.5 % for those with advanced degrees, quite possibly the result of random noise in the data. It is also conceivable that the fraction of people who understand how difficult a Mars mission would be, and how far we are from being ready to launch one, might be greater among those with the highest level of education.

Two American Mars missions failed late in 1999, the Mars Climate Orbiter on September 23 and the Mars Polar Lander on December 3. Based on a poll of 1,037

American adults, Gallup suggested that these failures were responsible for some public loss of confidence in NASA: "As NASA engineers search for the cause of problems that doomed the space agency's latest attempts to study the surface of Mars, a new CNN/USA Today/Gallup poll finds the American public's confidence in NASA has slipped. Slightly over half of those polled (53 %) between December 9–12, 1999 rate the job NASA is doing as either excellent or good, 53 %, while the percentage who rate it more negatively, as either fair or poor, has risen to 43 %. By way of comparison, when last asked this question in July—coinciding with the 30th anniversary of the Apollo 11 lunar landing—64 % of Americans rated NASA's performance in positive terms, and only 25 % believed it was doing less than a good job (Gillespie 1999)."

Under these circumstances, the poll asked "Do you think the federal government should or should not continue to fund efforts by NASA to send unmanned missions to explore the planet Mars?" In a subset of 513 respondents who were asked a question about unmanned Mars missions, 56 % felt funding for them should continue, versus 40 % who wanted them ended. Consistent with what we have found in many datasets, Gallup reported a difference by level of education. Among college graduates, 75 % wanted these missions to continue, versus 42 % of those who had not gone beyond high school.

In a 2001 report that reprised the results of that 1999 poll, Gallup reprinted figures from nine polls beginning in 1990 that asked how good a job NASA was doing: "The low point for America's space agency came in September 1993, after the loss of the $1 billion Mars Observer, when just 43 % of the public thought they were doing an excellent or good job (Carlson 2001)." Apparently the public was paying attention to news from the red planet, even though astronauts were not involved. In 2004, immediately after the successful landings of Spirit and Opportunity on Mars, Gallup reported that 70 % of 1,001 American adults polled felt this was "a major achievement," while 20 % called it "a minor achievement," and a tiny 9 % denied it was an achievement at all (Welch 2004).

Gallup reports that the level of popular support for a manned Mars expedition has remained rather constant over time. In July 1969, at the time of the first Moon landing, 39 % were in favor of sending an astronaut to Mars, compared with 43 % in July 1999 and 40 % in June 2005. The percent lacking an opinion was always small, but dropped slightly, from 8 % to 3 % to 2 %. So the percent definitely opposed to a Mars expedition was always the majority, from 53 % to 54 % to 58 %. In reporting these results, Gallup examined two demographic factors shaping them: "Men and women take very different views of funding a Mars landing. Men favor funding such a mission by 51 % to 47 %, but women oppose it by 68 % to 29 %. Additionally, 47 % of adults under age 50 support the United States undertaking a mission to Mars, compared with just 31 % of those aged 50 and older (Jones 2005)."

In February 2013, with support from the Boeing aerospace company, a poll called "Mars Generation" asked about the steps toward exploration of the red planet, as well as the goals it could serve. Its report said that for the 1,101 American respondents, "the top three reasons for human exploration of Mars are

(1) to achieve a greater understanding of Mars, (2) to search for signs of life, and (3) to maintain U.S. leadership in commercial, scientific and national defense applications (Mars Generation National Opinion Poll 2013)." These are plausible meanings Mars research may have for the public, but I suggest that the findings from this dataset be used with the understanding that it does not really represent a random sample of the population, and thus serves better as a source of concept categories than as a measure of the popularity of each.

The report was written with great enthusiasm from the standpoint of a group within the spaceflight social movement called Explore Mars, which says it was "created to advance the goal of sending humans to Mars within the next two decades," with the more general mission to "make humans a multi-planet species." The questionnaire was administered over Internet via email, using some plausible quality control techniques, but undoubtedly oversampling people more likely to have favorable opinions. Because this was not an ethnographic questionnaire, and the somewhat small number of items were written after the style of traditional opinion research, its ability to look deeply into respondents' conceptions is limited. However, some items did focus usefully on the chief alternatives under consideration. Here are the percentages who strongly agreed with four statements:

25.9 % Returning to the Moon is necessary before sending humans to explore Mars
20.6 % Human exploration of an asteroid would be worthwhile
21.0 % I am confident humans will go to Mars by 2033
20.7 % I am confident humans will go to Mars in my lifetime

Substantial numbers, from 41 % to 59 %, agreed with these items, but not strongly. A multiple-choice question asked respondents to identify the rover that landed on Mars in 2012, among these choices: Spirit, Curiosity, Opportunity, Viking I. The correct answer, Curiosity, was selected by 53.4 %. This suggests that respondents were especially attentive, although chance responding could account for nearly half of this figure.

5.5 Going to Mars Oneself

If spaceflight is a great human adventure, must it be experienced only vicariously, or can a young person realistically hope to travel to Mars at some time in the future? Asking respondents how they feel about going to Mars personally can help illuminate what the concept means to them. The 1986 questionnaire study of Harvard students, that was the focus of Chap. 4, included two questions about Mars. They were in the first of the pair of questionnaires, S1986A, which was primarily designed to elicit ideas about the goals of the space program through a few open-ended questions where respondents wrote in their answers. But the questionnaire also included many fixed-choice questions, several of which will be analyzed here. As in the case of S1986B, a subset of the data was published in an educational software system, based on responses from equal numbers of male and female respondents who had answered a large number of the questions properly, totaling 512 cases (Bainbridge 1992).

The first Mars question came after five other space questions and began with a prefatory sentence: "There has been much discussion about attempting to land people on the planet Mars. How would you feel about such an attempt—would you favor or oppose the United States setting aside money for such a project?" Of the 512 Harvard students, 244 or 48.4 % favored the US investing in a Mars expedition. A smaller faction, 173 students or 34.3 %, opposed it, 87 or 17.3 % had no opinion, and 8 students did not answer.

The second Mars question asked: "If you were asked to go along on the first rocket trip to the planet Mars, would you want to go or not?" A majority of the respondents, 55.9 %, said yes; 38.7 % said no, and 5.3 % had no opinion. Of course this is a hypothetical question, and only a tiny fraction might actually sign up for such a trip if it were offered to them. But it does measure a personal attitude about such a trip which can help us understand what it means to some people, especially when responses are correlated with those to other items in the questionnaire.

A set of 68 items in S1986A were preference questions, beginning with a list of 32 academic subjects: "Following is a list of various subjects that are taught at universities. Please tell us how much you like each one of them, whether you actually have taken a course in it or not." Answers were recorded for each one on a 7-point scale from 1 = do not like to 7 = like very much. The remaining 36 preference questions were more varied in topic area. Shortly after these 68 came a standard question about political orientation, asking respondents to describe themselves on a similar 7-point scale from 1 = extremely liberal to 7 = extremely conservative. Here we shall consider it along with the preference questions, and indeed it could be reconceptualized as asking the respondent's political preference in terms of the degree of liking conservatism. Table 5.4 analyzes some of the most relevant of these 69 items in terms of the two about Mars.

Analyzing the first item in the table, astronomy, will explain what the figures mean. Gamma is one of several correlation coefficients that would be appropriate for these data, summarizing the relationship between the two variables, one of which has two values (oppose or favor) that represent categories, and the other of which has 7 numerical values of the astronomy preference scale. Students who had no opinion or did not answer were excluded from the analysis. For this relatively small number of respondents, 0.18 just barely reaches statistical significance, but it seems logical to assume it represents a real but weak relationship between liking astronomy and favoring a Mars project. Among those who oppose funding a Mars project, the mean preference rating for astronomy on the 7-point scale is 4.28. As the positive gamma suggests, the mean preference for astronomy among those favoring the project would be a bit higher, and it is 4.67. The gamma between liking astronomy and wanting to go to Mars is higher, 0.29 compared with 0.18, and the difference in the mean preference scores is greater.

It might seem obvious that liking astronomy would correlate with positive answers to both Mars questions, but it might be surprising that the correlation is

Table 5.4 Two Mars questions and preference ratings for related topics

Preference for	US set aside money for Mars project			Respondent wants to go to Mars		
	Gamma	Oppose (173)	Favor (244)	Gamma	No (196)	Yes (283)
Astronomy	0.18	4.28	4.67	0.29	4.08	4.72
Physics	0.06	3.53	3.73	0.19	3.34	3.89
Engineering	0.19	2.88	3.39	0.26	2.83	3.45
The sciences, in general	0.15	4.46	4.86	0.16	4.46	4.88
Science fiction	0.28	4.01	4.82	0.38	3.83	4.92
Factual science articles	0.08	3.81	4.04	0.07	3.81	4.00
Driving very fast in a car	0.20	4.29	4.89	0.25	4.15	4.89
Taking physical risks	0.22	4.17	4.72	0.33	4.03	4.84
Complete personal security	−0.12	4.82	4.54	−0.22	4.95	4.43
Political conservatism	0.28	2.89	3.47	0.06	3.17	3.32

much stronger with personally wanting to go there. We see a comparable difference for liking physics, but with lower gammas. Indeed, the 0.06 between preference for physics and wanting the US to set money aside for a Mars project is statistically insignificant. The results for engineering are almost identical to those for astronomy, indicating that the Mars project would be an engineering effort as well as an astronomical one. The pattern for "the sciences, in general," is different, essentially the same for both Mars items, showing a somewhat weak correlation for both (0.15 and 0.16).

We would need new data, both larger numbers of respondents and additional questions, to be able to establish with great certainty what is going on here, but the following interpretation is plausible. Despite the pro-space rhetoric that an expedition to Mars would advance science in general, enthusiasm for it varies noticeably across people interested in different fields which we might imagine were relevant to the project. The greatest enthusiasm comes from people who in some way psychologically identify personally with the exploration of Mars.

The question about preference for science fiction was part of a set about types of literature, and shows both stronger correlations than those for astronomy and engineering, but the same pattern across the two Mars items. Yes, science fiction fans want an expedition to be funded (gamma = 0.28), but they more strongly wish they could participate personally (0.38). Three later chapters of this book will be devoted to science fiction, as it is manifested in different mass media, but for now we can note the relative strength of the connection to attitudes about real spaceflight.

The next three items concern attitudes taking risks in real life, which logically would predict willingness to contemplate flying to Mars. Diving very fast in a car risks crashes less spectacular but perhaps no less fatal than those a spacecraft might experience in a faulty Mars landing, and the entire flight would involve physical risks. A preference for "complete personal security" has negative correlations with the two Mars items. One could explain much of the results for the three risk items in terms of gender differences, given the greater male support for the space program, but doing so would not deny the risk-Mars patterns, but merely seek to determine the origins of risk preferences. The statistically significant gammas (0.20 and 0.22) connecting the first two risk items with support for the Mars expedition, quite apart from whether one imagined participating in it personally, could be the spurious result of gender differences. Or, they could reflect connections between attitudes about personal risk and about abstract risk experienced by someone else who cannot now be identified.

The results for the conservatism item are strikingly different, a stronger gamma with support for Mars investment than with going oneself, 0.28 versus 0.06. This is the opposite of the astronomy, engineering, and science fiction patterns. Earlier we saw some evidence that this same liberal-conservative scale might have a curvilinear relationship to support for the space program, with both liberals and conservatives being more supportive than moderates. Given how liberal Harvard students tend to be, and how much they discuss social issues and are aware of radical philosophies that are critical of the American system, it may be that Harvard liberals are especially conscious that funding for the space program may reduce funding for social programs. But if a Mars expedition were already happening, liberals and conservatives do not differ much on whether they would want to go to the red planet themselves.

Social science would be much easier if reality were more regular. Then all gamma correlations might be either -1.00, 0.00 or $+1.00$, reflecting a perfect negative relationship, no relationship at all, or a perfect positive relationship between two variables. But three main problems complicate the picture. First, causes operate in competition with each other, thus reducing each other's correlations with effects. Second, questionnaire items do an imperfect job measuring underlying realities. Third, the cost of large samples of respondents, and the impossibility of achieving true random samples of the population, add noise to the data. Thus, there is a tendency for social scientists to focus on statistically significant linear relationships, rather than contemplating what the full complexity of the data might mean. A simple example is afforded by examining the relationship between the two Mars questions.

Leaving out those having no opinion, 398 Harvard students gave clear answers to both Mars questions, producing the equivalent of a 2×2 table. The gamma correlation between the two Mars items is huge, 0.65, and for those readers who prefer the often more modest tau-b or Pearson's r, they are both 0.36. Of these 398 respondents, 177 both supported investment in a Mars project, and wanted to go along on the first expedition. At the other extreme, 99 opposed the project and would not go. These are both consistent positions.

An additional 59 Harvard students who favored the project but would not themselves go on the expedition. Although a little more complex, this also could be consistent, presumably if these respondents were disabled, in poor health, or especially concerned about avoiding physical risks. Perhaps a few favor general scientific and technological progress, but are personally involved in very different activities and careers which they would not interrupt for a Mars journey. I wonder if they also include a few visionaries or technically sophisticated students who believe that rockets—mentioned in the question—are not the right means for a Mars expedition.

Once those three categories are accounted for, another 63 students were opposed to US investment in a Mars project, but would want to go on the first expedition, a seeming contradiction. Perhaps, as suggested above, some are political liberals who want the money spent on social programs, but if a Mars expedition were in progress they might as well join it. Some might believe that the first Mars expedition will be carried out by a private company, an international coalition, or even a mad scientist with the genius to invent a new kind of rocket ship, like Dr. Zarkov from the *Flash Gordon* stories. Some may be technological realists who believe that any money invested in a Mars project today would be wasted, because additional advances in pure science and advanced engineering would be required before the first expedition would really be feasible. A respondent who was 20 years old when answering S1986A could well imagine going on the expedition at age 55, but that would mean going in 2011, and that year, sad to say, is already behind us and no Mars expedition is even on the horizon.

5.6 Adolescent Attitudes

A dozen years after the Harvard study, I had the opportunity to help design "Survey2000," a web-based questionnaire focusing on migration and regional culture, sponsored by the National Geographic Society and organized by sociologist James Witte (Witte et al. 2000; Witte and Pargas 2004). From late September through early November 1998, more than 50,000 adults and 15,000 children responded to this half-hour online survey. Although the "National" in "National Geographic Society" refers to the United States, this was a truly world-wide study. Four nations contributed more than 1,000 adult respondents each: Australia, Canada, the United Kingdom, and the United States. Thirty-one others contributed more than 100 each: Argentina, Belgium, Brazil, China, Chile, Colombia, Denmark, Finland, France, Germany, Greece, Hungary, India, Ireland, Israel, Italy, Japan, Malaysia, Mexico, Netherlands, New Zealand, Norway, Philippines, Portugal, Singapore, South Africa, Spain, Sweden, Switzerland. Turkey, and Venezuela.

The project was remarkable not only for being done cost-effectively over the World Wide Web, but for being one of the most technically complex questionnaire studies ever carried out by any means, exploiting many of the exciting new capabilities of computer-based electronic communications. Early questions tracked the US and Canadian adult respondents' residential moves, and the computer selected later items on the basis of the individual's history of geographic mobility, providing data used in my recent book *Personality Capture and Emulation*. Questions about preferences in food and literature were automatically tailored to the regions of birth and current residence, and one battery of music preference items actually let the respondent hear samples of the music. Other questions concerned the respondent's Internet use versus involvement in the local community.

My main initial interest was the youth survey, especially the version administered to children aged thirteen to fifteen (Bainbridge 2000, 2002). Other researchers had already adapted the section of the adult survey concerning preferences in music and had developed a long section asking about the child's favorite activities and interests. In addition to consulting on the overall content of the youth survey, I added items about science and technology. Each was a statement, such as "There should be a law against cloning human beings," and the respondent was supposed to respond either "strongly agree," "agree," "disagree," or "strongly disagree."

One was a variant of the key item from the Harvard study: "If I were asked to go along on the first rocket trip to Mars, I would go." Near a third of the 3,185 children who responded, 32.5 %, strongly agreed. We should not conclude that they are really ready to step aboard a spaceship to Mars, with or without their parents' permission. Rather, this is a measure of a generally positive orientation toward the idea. Another 28.9 % agreed but not strongly, while 23.6 % disagreed, and 15.0 % disagreed strongly.

As we might expect from many of the adult polls we have considered to this point, there was a significant gender difference. Among 1,461 boys, 39.9 % agreed, compared with 26.0 % of 1,671 girls. Given the large numbers of both boys and girls who gave each possible response, we are in a good position to learn more about the gender differences in the following analysis.

The youth version of Survey 2000 contained 53 questions about the child's favorite activities or interests, from acting in dramas to watching TV. Really this set of items was just a list with an HTML checkbox next to each, so the respondent could use the computer's mouse to check off his or her favorites. Of these 53, there were 19 that showed some statistical connection with wanting to go to Mars, and they are listed in Table 5.5. Of those respondents who clicked the box next to "Astronomy," indicating it was one of their favorites, 47.1 % said they wanted to go to Mars, compared with 27.0 % of those who did not click the box. This is a difference of' 20.1 % points, and I have arranged the 19 items in Table 5.5 in descending order of this difference.

The inclusion of some of the activities and interests listed in Table 5.5 is a mystery, but several make perfect sense. Certainly going to Mars would be an adventure in practical astronomy, contributing to science in general, but it would take unusual optimism to believe one would really do archaeology there. Several of the activities would be required during the mission, including rock climbing, hiking, and exotic versions of camping and being a boy scout or girl scout. The organized sports listed in Table 5.5 have nothing to do with Mars, but the Harvard study implied these items may primarily reflect willingness to take physical risks which one would surely do on the first rocket trip to Mars. On the other hand, perhaps many items merely distinguish the boys from the girls.

Consider video games. For all the children, knowing whether a child likes to play video games improves your ability to predict whether he or she is enthusiastic

Table 5.5 Favorite activities or interests and percent who want to go to Mars

Activity	Percent who strongly agree they would go to Mars		Percentage difference (%)
	A favorite activity (%)	Not a favorite activity (%)	
Astronomy	47.1	27.0	20.1
Archaeology	42.6	30.2	12.4
Science	39.6	28.0	11.6
Rock climbing	41.0	29.5	11.5
Martial arts	42.3	30.9	11.4
Scouts or guides	41.9	31.4	10.5
History	39.3	29.2	10.1
Sailing	39.2	31.2	8.0
Geography	38.3	30.3	8.0
Mathematics	38.0	30.1	7.9
Computers	35.2	27.9	7.3
Skateboarding	38.7	31.2	7.5
U.S. football	37.5	30.1	7.4
Video Games	35.3	29.2	6.1
Camping	35.8	29.8	6.0
Skiing	36.5	30.8	5.7
Hiking	36.1	30.8	5.3
Fishing	36.0	30.7	5.3
Soccer football	34.9	30.8	4.1

about going to Mars. A higher proportion of video-game players are than of non-players. But this is entirely a reflection of the fact that boys tend to like video games. When I analyzed the boys and girls separately, there was no connection whatsoever for video game players of either sex to want to go to Mars more than non-players. The situation is quite different for skateboarding. Among boys, preference for skateboarding has absolutely no connection with wanting to go to Mars. But among girls, there is a big connection. Skateboarding may be a better measure of willingness to take physical risks for girls, few of whom skateboard, than among boys, where many do.

This is not the place to go into a deep technical analysis of the statistics, but it is worth taking a quick look. The gamma correlation between liking astronomy and wanting to go to Mars is 0.35, a very solid positive correlation. If we separate the two sexes, gamma drops for boys to 0.28 and rises for girls to 0.39. For all students, there is a gamma of 0.20 between liking martial arts and wanting to go to Mars, but the correlation is only 0.12 for boys and fully 0.23 for girls. The overall skateboarding correlation is 0.14, but just 0.01 for boys and fully 0.22 for girls. These are actually very important findings, because they show that some factors which distinguish attitudes about going to Mars work more powerfully for girls than for boys.

Three of the more violent male sports—skateboarding, U.S. football, and soccer football—give extremely revealing results. For children in general, preferences for these three sports are significantly correlated with interest in going to Mars. Among girls, there are somewhat higher correlations. But among boys, the correlations are dead zero—no relationship at all. Nothing logically connects these sports with being a future Martian; they are merely good measures of masculinity. From this we might conclude that girls who want to go to Mars are "tomboys," young females with boyish habits. But that would be inaccurate. Rather, some girls may have a wide range of interests, combining traditionally feminine interests with those that are traditionally masculine. Among girls, being a potential Martian is associated with a degree of liberation from traditional gender stereotypes, just as becoming Martian would be liberation from Earthbound conventions.

Among thirteen- to fifteen-year-old respondents to Survey2000, boys do show somewhat more support for science than do girls. The Mars trip item was one of twelve that presented respondents with a statement and asked how much they agreed or disagreed with it. The first said, "Science will do more good than harm in the next century." Among boys, 35.5 % strongly agreed with this statement, compared with 21.8 % of girls. Two items focused on general support for the space program. More boys, 24.2 % compared with 11.8 %, strongly agreed that "Funding for the space program should be increased." Almost exactly equal percentages, 12.8 % of boys and 13.0 % of girls, strongly agreed that "Space exploration should be delayed until we have solved more of our problems here on earth." But a big gender difference shows up when we look at disagreement. Fully 27.0 % of boys strongly disagreed with the idea that space exploration should be delayed, compared with only 14.3 % of girls. On many of these items, boys show more of a tendency to strongly agree or strongly disagree, whereas girls merely agree or disagree. But looking past this tendency of boys to give more extreme answers, we find that boys also tend to agree with pro-science statements, and to disagree with anti-science statements.

In Table 5.6 we see that interest in Mars tends to be found not only among those who want funding for the space program to be increased, but also among those who support research on human cloning, who don't worry much about environmental problems, who are generally optimistic about the future, and who think development of nuclear power should continue. In contrast, there is less interest on average among those who believe there should be a law against human cloning and who want space exploration delayed until terrestrial problems have been solved. Thus, science issues not directly related to Mars, notably attitudes toward human cloning, predict Martian attitudes.

Interestingly, I did not find the same pattern within the genders for science attitudes as I did for interests and activities. The correlations between wanting to go to Mars and each of the items in Table 5.6 are almost identical among boys and girls. Even though interest in science is greater among boys than among girls, it is not a measure of masculine culture that distinguishes cosmopolitan girls from non-cosmopolitan girls. This suggests that girls can remain "feminine" while becoming scientific. To put this another way, women may enter scientific fields and achieve within them without having to possess unusual personalities or radical values.

Table 5.6 Science attitudes and percent who want to go to Mars

Statement	Strongly agree (%)	Strongly disagree (%)	Percentage difference
Funding for the space program should be increased	62.40	24.90	37.5
Research on human cloning should be encouraged, because it will have great benefits for science and medicine	57.40	25.90	31.5
We should not worry much about environmental problems, because modem science will solve them with little change to our way of life	56.50	33.60	22.9
All in all, the world's population will be better off in the next 100 years	54.40	31.60	22.8
Development of nuclear power should continue, because the benefits strongly outweigh the harmful results	55.50	33.40	22.1
Science will do more good than harm in the next century	49.70	40.60	9.1
We should accept cuts in our standard of living in order to protect the environment	42.60	38.50	4.1
All nuclear power plants should be shut down or converted to safer fuels	36.90	40.30	−3.4
Intelligent life probably does not exist on any planet but our own	38.20	44.60	−6.4
There should be a law against cloning human beings	32.20	45.90	−13.7
Space exploration should be delayed until we have solved more of our problems here on Earth	32.40	54.40	−22.0

Early in 2004, when the successful landings of the unmanned Mars rovers, Spirit and Opportunity, were in the news, an online Gallup survey of 785 American youth aged 13 to 17 included two agree-disagree items about personal spaceflight: "If

Mondak, J. J. (1993). Source cues and policy approval: The cognitive dynamics of public support for the Reagan agenda. *American Journal of Political Science, 37*(1), 186–212.

National Research Council (2012). *NASA's strategic direction and the need for a national consensus* (p. 2). Washington DC: National Academies Press.

Pion, G. M., & Lipsey, M. W. (1981). Public attitudes toward science and technology: What have the surveys told us? *The Public Opinion Quarterly*, *45*(3): 303–316.

Russett, B. (1990–1991). Doves, hawks, and U.S. public opinion. *Political Science Quarterly, 105*(4), 515–538.

Slater, J., & Goldfischer, D. (1986). Can SDI provide a defense. *Political Science Quarterly, 101*(5), 839–856.

Smith, T. W. (1988). Nuclear anxiety. *The. Public Opinion Quarterly, 52*(4), 557–575.

Stine, D. D. (2009). U.S. civilian space priorities: Reflections 50 years after Sputnik, Congressional Research Service report for Congress, (p. 13). Retrieved February 26, 2014, from http://fpc.state.gov/documents/organization/108077.pdf.

Vaughan, D. (1990). Autonomy, interdependence, and social control: NASA and the space shuttle challenger. *Administrative Science Quarterly, 35*(2), 225–257.

von Braun, W. (1953). *The Mars project.* Urbana: University of Illinois Press.

Welch, D. (2004). Billions and billions? As long as it's not dollars, press release February 24, Gallup. Retrieved November 10, 2013, from http://www.gallup.com/poll/10777/Billions-Billions-Long-Its-Dollars.aspx.

Wells, H. G. (1898). *The war of the worlds*. New York: Harper.

Witte, J., & Pargas, R. (2004). Online questionnaires. In W. S. Bainbridge (Ed.), *Berkshire encyclopedia of human computer interaction* (pp. 520–525). Great Barrington: Berkshire Publishing Group.

Witte, J., Amoroso, L., & Howard, P. N. (2000). Method and representation in internet-based survey tools: Mobility, community, and cultural identity in Survey 2000. *Social Science Computer Review, 18*(2), 179–195.

Zakaria, F. (1990). The Reagan strategy of containment. *Political Science Quarterly, 105*(3), 373–395.

Zubrin, R. (1996). *The case for Mars*. New York: Free Press.

Chapter 6
Sciences

Many academic questionnaire studies and public opinion polls allow us to examine how attitudes about space exploration correlate with interest and support for many specific sciences. This chapter will therefore be able to place space exploration within the general context of society's conceptions of the sciences. Naturally, some sciences like physics are directly connected, others like geology indirectly, and some like archaeology perhaps not at all. Significantly, science is divided into very distinct cultures, starting with the physical, biological, and social sciences. One can conceptualize the exploration of Mars in terms of astronomy or geology, and these two sciences are very different in their methods and intellectual frameworks. Factors such as the level of education achieved by questionnaire respondents affect not only their general attitudes, but also how finely differentiated their opinions are.

Certainly, space exploration contributes to progress in several sciences, most obviously astronomy, but also physics through observing extreme phenomena that do not exist on Earth, and geology by providing comparison between Earth and other planets. Yet, the space program was not primarily or even significantly motivated by scientific curiosity, and academic science has been frankly ambivalent about it. Given current policy debates and the uncertain future of the manned space program, we can imagine a scenario in which the space program does become exclusively a platform for research in fundamental science, probably for the foreseeable future without human expeditions beyond low Earth orbit, yet this might cause opposition from the general public who may have defined the program's traditional emphasis as science.

6.1 Attitudes of Scientists

Perhaps the most significant poll of spaceflight attitudes done during the 1960s was not carried out by Harris or Gallup, nor did it involve the general public, but it was a survey of scientists conducted by the editors of that prestigious journal,

W.S. Bainbridge, *The Meaning and Value of Spaceflight*,
Space and Society, DOI 10.1007/978-3-319-07878-6_6

Science, and published in its July 24, 1964 issue. Here is how I summarized it in *The Spaceflight Revolution*, the book I wrote just a decade later:

> In 1964 *Science* mailed a questionnaire on space to 2000 members of the American Association for the Advancement of Science. Replies were received from 1134, including 548 with Ph.D. degrees. Only 20 percent of the whole sample and a mere 16 percent of the Ph.D.s felt that "a reasonable objective would be a lunar landing by... 1970," the target President Kennedy had set three years earlier. Sixty-two percent of the whole sample and 64 percent of the Ph.D.s disagreed that "the vital national interest of the United States require that a high priority be given to landing a man on the moon by 1970" (Bainbridge 1976a).

The question about the year set for the first lunar landing offered responses that were dates later than 1970, and only 7 % said that this should never be our goal. But later years were selected by most respondents: 23 % chose 1975, 19 % voted for 1980, and 9 % for 1990. When asked to select from a list "the most important justification for manned exploration of the moon," they rated science highest at 58 %, putting national prestige second at 13 %. But this does not mean they were very enthusiastic about the Apollo Program. Most significantly, when given a list of scientific fields and asked to identify the top one in terms of potential for new knowledge, this was their ranking:

51 % Biomedical research
15 % Physics research
10 % Manned lunar research
9 % Chemical research
7 % Other
6 % Oceanography
2 % Earth-based astronomy.

In NASA and the spaceflight movement more generally, the report of this poll was seen as an attack on the Apollo Program, and rightly so. I know from my own experience, as well as from studying history, that many scientists privately are very critical of the manner in which the US government decides which science-related projects to fund or encourage. In particular, the Kennedy-Johnson administration decided about the Apollo Program behind closed doors, and did not poll the wider scientific community about which sciences were ready for rapid advancement if increased funding for them were available. In reaction, advocates of the program claimed that the AAAS poll was unscientific. For example, in its annual volume summarizing space news, the NASA policy office devoted two paragraphs to this poll, one summarizing the results, and this second paragraph that sought to refute them:

> Commenting on the poll, Robert C. Toth of *Los Angeles Times* noted that *Science* editor Philip Abelson was "an outspoken opponent of the moon program" and that the survey did not directly ask the key question: Should the existing program be slowed down now? *Journal of the Armed Forces* space editor James J. Haggerty, Jr., pointed out that "it all depends on who you ask." Of the 1,134 respondents, he said, "it turned out that 86 percent were not connected with the space program. As a matter of fact, there is a strong suspicion that they are all medical researchers, since 51 percent of them indicated that biomedical research is the scientific field with the potential for producing the greatest amount

of new knowledge, where only 10 percent thought manned lunar research would contribute anything of value" (NASA Historical Staff, Office of Policy Planning, Scientific and Technical Information Division 1964).

One would have thought that the proper response to this debate would have been a new poll, conducted by social scientists with experience in survey research, perhaps a whole series of careful scientific studies of how NASA and its Moon program fit into science more generally, and considering which other scientific areas might have the potential to achieve more for the same dollar investment. This was never done, and there were several reasons, of which four deserve mention.

As a sociologist, with considerable experience in survey research, the first thing I note is that *Science* does not consider sociology to be a science. In the 1964 poll about the Moon program, all the social and behavioral sciences are crammed into the Other category. The journal does not solicit articles by sociologists, and behavioral science appears almost exclusively when the research examines physical data, as in neurobiology. The journal *Nature* does not publish much if any social science either, but it has the excuse that its title concerns the natural world, not culture. I am happy to report that I myself have published in *Science*, but all but two of my contributions were book reviews, one of them space-related (Bainbridge 1998). One non-review was an op-ed piece about the need for research about intellectual property issues in the Age of Internet (Bainbridge 2003). The one full article I published in *Science* was about new computer technologies that had social implications (Bainbridge 2007). Ordinary sociology does not appear in that journal.

A second reason is that the federal government has never developed a policy or mechanism for allocating funding across the sciences in terms of their potential to achieve new discoveries. This point was forcefully argued by sociologist Amitai Etzioni in the pages of *Administrative Science Quarterly* back in 1966 (Etzioni 1966). His observation is factually correct but also has debatable policy implications, because like many sociologists Etzioni believes that sociology should become the primary guide for society in developing its institutions and dealing with its problems, while many non-sociologists consider this to be undesirable left-wing dogma. Earlier, Etzioni had written the book *TheMoon-Doggle*, the title of which suggests that the Apollo Program was a boondoggle benefiting aerospace companies more than science (Etzioni 1964). His Wikipedia page makes his analysis sound quite reasonable, calling it "an early reason-based critique of the space race… in which he points out that unmanned space exploration yields a vastly higher scientific result-per-expenditure than a manned space program".

The third reason is that many members of society confuse science with engineering. "It's not rocket science," we often say, yet there is no such thing as rocket science. Rather there is rocket engineering, which uses principles discovered in many fields of science, notably physics and chemistry, in the design and development of space propulsion machinery. A good comparison is computer science, which I argue does not exist, either. Rather, computers are created by electrical engineers, with input from applied mathematicians and people in other fields.

A Computer Science movement was able to convince first government and then universities that a well-delineated science of computers could exist, but at some point in the near future they will need to deal with the fact that electrical engineering is reaching its own natural limits, and the field will require redefinition or dissolution. Government support for "computer science" may have been a better investment than the space program, for example in the hugely important development of Internet. But the reason for the success is that small teams, like the "two guys in a garage" that gave us Apple computers, could make great contributions to information technology, which is not the case for spaceflight. If government seeks engineering solutions to its problems, but the term *science* has higher prestige than *engineering*, then all kinds of confusions and occasional bad decisions will ensue.

The fourth reason is actually a complex system of factors, starting from the point that scientists in any one field tend to form a community in competition with others for support from corporations and government agencies. At the present time, social science seems to have very little to offer industrial corporations, or the military, and only economists are widely employed in either context. I recall not many years ago attending a lecture by a leader associated with the Webb Space Telescope, who extolled the virtues of this new instrument that would study the universe at infrared wavelengths. Despite being enthusiastic about it myself, I had to ask the question: "Is this the right priority for science, given that the same amount of funding could double the budget of the NSF Sociology Program for hundreds of years." The lecturer's response was that we must not think of scientific investments as being in competition with each other. The subsequent cost overruns that have brought the total cost of the Webb to $8 billion would not double the Sociology Program for hundreds of years, but for 1,000 years.

In 2004, James van Allen called for dispassionate consideration of whether manned spaceflight was worth the costs. The headline for his essay was a rhetorical question: "Risk is high, cost is enormous, science is insignificant. Does anyone have a good rationale for sending humans into space?" His answer was no. Van Allen was not some mere, biased sociologist, but the space scientist after whom the van Allen radiation belts were named. His conclusion drew an analogy between human spaceflight and the high-altitude balloon flights that in the 1930s promised great scientific returns but achieved hardly any: "Have we now reached the point where human spaceflight is also obsolete? I submit this question for thoughtful consideration. Let us not obfuscate the issue with false analogies to Christopher Columbus, Ferdinand Magellan, and Lewis and Clark, or with visions of establishing a pleasant tourist resort on the planet Mars" (van Allen 2004).

Spaceflight historian Roger Launius has disputed van Allen's analysis, arguing that reducing funding for the manned space program would not free up funds for unmanned programs or other more scientific missions (Launius 2006). Once the US manned space program had begun, closing it down would incur political costs, even if only a minority of voters really supported it, possibly great enough to decide the outcome of the next election if the party out of power vigorously attacked the administration that cancelled it, using empty rhetoric about scientific value as well as national prestige. Given that the money comes from taxpayers, and the US government has never paid off all its debts since the Second World War, it is very difficult to campaign for entirely new programs of scientific research, without help from an already influential interest group.

6.2 The Nature of Science

People differ greatly in how they conceptualize science, having different views on what it is and about what value it has for humanity. Thus they differ in how they connect science with spaceflight, and what priority they give this connection in making decisions about which programs or projects deserve public investments. A wealth of science-related poll data exist, although seldom designed to answer the deepest questions. Therefore, we need a theoretical perspective to use as a starting point, without necessarily being able to test it empirically.

The word *science* is derived from the Latin *scientia*, which roughly means *knowledge*. It is said that the English word with the largest number of different meanings is *run*, yet *know* has many as well, referring to personal intimacy in many senses as well as simply possession of facts. The same was true in Latin, where the root of *scientia* is *scire*, translated alternately into English as "to know, understand, perceive, have knowledge of, be skilled in" (Lewis 1916). The first person singular active form is *scio*, meaning *I know* in all these senses, and modern students of Latin are expected to memorize these four tense root forms: *scio, scire, scivi, scitum*. However, there is a joke form that some students learn, based on an English pun with *ski*: *scio, slipperi, falli, bumptum*. This is not merely humor but philosophy, suggesting that the pursuit of knowledge can go badly awry and lead to disaster.

By using a word derived from Latin to name science, we abstract it from our everyday experience, thereby introducing uncertainty about what it really means. It is useful to conceptualize science in terms of two competing but potentially compatible theories, each stating perhaps too categorically the ultimate goal of scientific work:

Technical: Scientists discover facts about nature that gain value for humanity when they are used by engineers to develop new technologies.

Philosophical: Scientists seek the truth so that humans will no longer be misled by superstitions and false impressions, but understand reality as it truly is.

In the early years of the twenty-first century, both of these meanings confront challenges, which could be the subjects of two other books, but must be mentioned briefly here because they relate to the meaning of spaceflight. The *technical* meaning of science is especially problematic, because it leaves open the question of whether discoveries will continue to facilitate new technologies. We may have reached the point at which we already know most of what there is to learn about nature, so additional discoveries will not be forthcoming, and technological innovation will slow to a halt. One example is the debate over whether controlled nuclear fusion for power production, or conceivably for rocket propulsion, will ever be achieved.

The *philosophical* meaning of science implies that it is good for ordinary people to understand that the Earth is not flat but round, that the Earth goes around the Sun rather than the reverse, and that the planets are not other Earths but large objects having their own distinctive and frankly rather unfriendly characteristics. This last point links to the fact that knowing a friend is different from knowing a fact, because interpersonal knowledge involves an emotional bond and a social relationship, rather than merely possession of cold facts. One significant implication of the philosophical theory is that science is likely to contradict many of our traditionally treasured beliefs, such as in religion, psychology, and politics. Relevant to this book is the issue that the social sciences may be especially subversive of popular beliefs, as seen in the frequent political attacks against federal funding of social science, and in the apparent failure of social science to make discoveries heralded in the popular news media. Since this book is itself social science, and may occasionally report results unfavorable to spaceflight, the low status of these sciences of human behavior may undercut the value of this very effort.

In the period 2006–2010, the General Social Survey included a number of questions about science and technology, many of which will feature in this chapter. One battery of items listed eight fields taught in universities, asking: "How scientific are each of the following fields? If you have not heard of a particular field, just say you haven't heard of it." Here are the fractions calling each field "very scientific," with the number of respondents including those who had not heard of it in parentheses:

82.3 % Medicine (2,312)
71.6 % Biology (2,288)
70.7 % Physics (2,254)
48.1 % Engineering (2,270)
16.6 % Economics (2,261)
13.4 % Accounting (2,755)
11.3 % History (2,282)
9.6 % Sociology (2,209).

The fact that sociology comes out last in this scientific sweepstakes does not mean the general public rejects it altogether, because 41.8 % called it "pretty scientific." But 31.4 % said it was "not too scientific," 8.3 % said it was "not scientific at all," and 8.8 % professed never to have heard of it. One might think that physics would be stereotyped as most scientific, but biology edges it out, and medicine heads the list. This is perplexing, because physics is often considered the most rigorous, even mathematical of the sciences, while biology has not yet shifted over to being primarily rigorous studies of DNA structures and other methods that compete directly with physics. Indeed, the high "scientific" status of medicine in the public mind is not new, and in a 1959 summary of poll data Stephen Withey observed, "for the public the caduceus of medicine sits proudly at the top of the totem pole of science" (Whithey 1959). It may be that people read personal relevance into "scientific," and medicine is the field of application where a science touches them most personally.

Astronomy was not in the list, so physics may stand in for its philosophical relevance to spaceflight, and engineering for its technical relevance. In any case, 48.1 % of respondents considered engineering to be "very scientific." Another 33.7 % considered it "pretty scientific," 11.5 % "not too scientific," 6.4 % "not scientific at all," and an infinitesimal 0.4 % had never heard of it. It may be that few people make a very clear distinction between science and engineering, and thus may consider technology advances in the space program to constitute scientific progress, rather than resulting from scientific progress achieved in other fields such as chemistry.

Government science agencies vigorously promote so-called *STEM* fields, an acronym that stands for Science, Technology, Engineering, and Mathematics. Taken literally, STEM implies science and mathematics are separate, flanking two industrial fields that can hardly be distinguished, technology and engineering. Indeed, mathematics cannot be a science, because it does not employ empirical methods, unless you count computerized searches for prime numbers, and similar highly peripheral activities. These same agencies, in the propaganda they use to seek public funding, stress the technical theory of science, evaluating it in terms of the worth of its technological products, rather than philosophically in terms of how it changes our view of existence.

We see this confusion between science and engineering also in reports on the budgets of government science agencies, but it is possible to disentangle them to some degree. The 2006–2010 GSS science module was largely funded to provide data for the biennial *Science and Engineering Indicators* reports of the National Science Board, compiled by the National Science Foundation. The latest such book-length report, published in 2012, breaks down research and development budgets in 2009 for US government agencies by three categories: basic research, applied research, and development which is primarily research-assisted engineering (National Science Board 2012).

Much of NASA's budget funds operations, such as the costs of building and operating launch vehicles, and in 2009 its research and development budget totaled $5,937,000,000. Of this, only 17.2 % was considered basic, 11.5 % was applied, and fully 71.3 % was development. In contrast, 92.3 % of the $6,095,200,000 R&D budget of the National Science Foundation was invested in basic research, only 7.7 % in applied research, and none in development. Here are the basic research percentages of the R&D budgets of the other agencies that spent over $1,000,000,000 in R&D: Department of Health and Human Services (52.8 %), Department of Energy (41.1 %), Department of Agriculture (40.7 %), Department of Commerce (12.1 %), and Department of Defense (2.5 %). By this comparison, NASA is an engineering agency, not a science agency.

NASA's 2014 budget request tabulated the actual expenditures for 2012 in the major budget categories. Including a small rescission, the total NASA budget for 2012 was $17,770,000,000. Of this, $5,073,700,000 was for science, somewhat less than the $5,937,000,000 for 2009, but that year many agencies had received extra American Recovery and Reinvestment Act funds. Within the science category, here are the percentages for the five subcategories defined by NASA: Earth Science (34.8 %), Planetary Science (29.6 %), Astrophysics (12.8 %), Heliophysics (12.7 %), and the James Webb Space Telescope (10.2 %). The request's message from NASA administrator Charles F. Bolden, who himself was an accomplished astronaut, summarized the agency's 2014 scientific goals:

> NASA's ground-breaking science missions are reaching farther into the solar system, revealing unknown aspects of the universe and providing critical data about our home planet and threats to it. Spacecraft are speeding to Jupiter, Pluto, and the dwarf planet Ceres, while satellites peer into other galaxies, spot planets around other stars, and work to uncover the origins of the universe. The budget funds an amazing fleet of scientific spacecraft. The budget request will also support our study of Earth and its response to natural or human-induced changes. On the heels of the most daring mission to Mars in history last year, this budget will provide funding to launch another mission to the Red Planet. We also will continue making steady progress to develop and conduct critical tests on the James Webb Space Telescope, leading to its planned launch in 2018. The telescope will revolutionize our understanding of the universe, just as its predecessor the Hubble Space Telescope did.

A sense of the public's conception of the space program can be derived from a standard series of thirteen General Social Survey items about confidence in societal institutions, as responded to by from 1,796 to 1,863 respondents in the 2008–2012 period. The GSS introduction to this battery of items reads: "I am going to name some institutions in this country. As far as the people running these institutions are concerned, would you say you have a great deal of confidence, only some confidence, or hardly any confidence at all in them?" Table 6.1 cross-tabulates with the GSS space funding item, arranging the institutions in descending order of their gamma correlations with funding.

Among respondents who have a great deal of confidence in the scientific community, 23.0 % feel that current space funding is too little, compared with only 8.4 % of those with hardy any confidence in the scientific community. Over the years, confidence in some institutions has varied greatly, and at this point in time the institution having the greatest public confidence is the military, 51.9 % having a great deal of confidence, with the scientific community in second place with 41.6 %. As measured by gamma, confidence in the military has a small positive correlation with wanting funding increased for space exploration, yet the table shows a curvilinear relationship, with somewhat more support for space exploration among respondents holding both extreme opinions, a great deal of confidence in the military or hardly any.

One can imagine many hypotheses to explain some of the other patterns in Table 6.1, for example conjecturing that confidence in the U.S. Supreme Court reflects higher than average education, and indeed college graduates have more confidence in the Court than other respondents, 35.3 % versus 27.7 %. But the really interesting thing we see is that education is at the very bottom, with just 13.1 % of those with a great deal of confidence in education feeling space funding is too little, compared with 22.4 % of those with hardly any confidence in education.

Table 6.1 Confidence in American intuitions and support for space funding

Institution	Great deal of confidence (%)	Space funding is too little			Gamma
		Great deal (%)	Only some (%)	Hardly any (%)	
Scientific community	41.6	23.0	14.1	8.4	0.31
U.S. supreme court	29.1	18.2	17.5	15.0	0.14
Major companies	14.9	17.9	16.6	18.1	0.12
Medicine	38.8	17.7	16.6	16.5	0.10
Military	51.9	17.7	15.6	18.8	0.09
Executive branch of the government	12.6	15.9	15.9	19.0	0.08
Organized religion	20.5	16.1	15.5	21.1	−0.01
Press	10.0	14.0	13.6	21.0	−0.04
Congress	7.3	11.9	14.3	20.8	−0.04
Organized labor	11.7	18.0	14.9	21.0	−0.05
Banks and financial institutions	12.3	11.4	14.7	22.6	−0.08
Television	11.2	11.1	15.7	20.2	−0.08
Education	25.8	13.1	17.1	22.4	−0.09

6.3 The 2006–2010 GSS Science Module

Education was a powerful variable in the GSS results that fed into *Science and Engineering Indicators*. For example, of 1,176 respondents who had never taken a college-level science course, just 9.5 % felt funding for space exploration was too little, compared with 21.3 % of the 889 who had taken at least one. Personal involvement with modern technology was a related factor and had a similar effect. Of the 657 respondents who did not have Internet connectivity at home, just 9.0 % felt space funding was too little, while 17.2 % of the 1,406 who had Internet at home wanted funding increased. These two factors were closely linked, having a gamma correlation of 0.66 and a tau-b correlation of 0.32, and reflected general social class differences. One could say that social class differences cause all the effects reported in this section, but that oversimplification would ignore the meaning and causes of social class differences themselves.

Education certainly reflects knowledge about fields sufficiently remote from daily life that without schooling an individual would not have much experience with them. One battery of knowledge items begins with this preamble: "Now, I would like to ask you a few short questions like those you might see on a television game show. For each statement that I read, please tell me if it is true or false. If you don't know or aren't sure, just tell me so, and we will skip to the next question. Remember true, false, or don't know. First, the center of the Earth is very hot. Is that true or false?" Table 6.2 gives the percentages answering correctly, plus

Table 6.2 Knowledge of science and support for space funding

Question	Answer	Percent correct (%)	Space funding too little		
			Answered correctly (%)	Answered incorrectly (%)	Gamma
The center of the earth is very hot	True	93.4	16.4	10.7	0.20
All radioactivity is man-made	False	81.0	17.4	11.0	0.27
It is the father's gene that decides whether the baby is a boy or a girl	True	73.2	16.5	14.6	0.02
Lasers work by focusing sound waves	False	69.8	20.1	9.8	0.26
Electrons are smaller than atoms	True	71.1	18.6	12.6	0.16
Antibiotics kill viruses as well as bacteria	False	58.8	18.4	10.6	0.26
The universe began with a huge explosion	True	50.4	20.2	9.0	0.37
The continents on which we live have been moving their locations for millions of years and will continue to move in the future	True	89.6	16.6	7.1	0.26
Human beings, as we know them today, developed from earlier species of animals	True	52.7	18.2	10.7	0.30

the percentages who feel space funding is too little among those answering correctly and incorrectly. The numbers of respondents vary from 1,404 to 1,981. The gamma correlation comes from a cross-tabulation with all three responses to the space funding item, and scored positively with giving a correct answer to the particular science knowledge question.

One item strikes me as poorly worded: "The universe began with a huge explosion." This clearly refers to the Big Bang, yet people really knowledgeable about the physics of the early universe may conceptualize the expansion not as an explosion, which implies the rapid expansion of matter into a pre-existing space, but rather as the generation of space itself. Indeed, among astrophysicists in recent decades there have been many debates about the possibly changing rates of spatial "expansion," including a possibly much greater early inflationary period, and today an acceleration (Guth 1981). But perhaps semantics is not at work here, but religion. Calling this statement true shows a very powerful negative correlation with church attendance, gamma = −0.44. Of 618 people who never attend church, 68.6 % judged this Big Bang item to be true, compared with only 16.4 % of the 262 who attend more than once a week, and all the other rates of church attendance show percentages ranging smoothly between these extremes. The item about human evolution from animals shows the same pattern, a gamma of −0.45

and a range calling this statement true from 69.7 % among those who never attend church, down to 15.7 % among those who attend more often than once a week.

With these two items, at least, we may be dealing not with knowledge but belief. Probably most respondents know what science says, but some refuse to accept it. The question about confidence in the scientific community shows a somewhat different pattern. Rather than responses varying smoothly from those who never attend to church to those who very frequently do, confidence in the scientific community is high across all levels of relatively infrequent attendance, but drops at the highest levels of attendance. For example 44.7 % of the 1,014 respondents in 2006–2010 who never attend say they have a great deal of confidence in the scientific community, compared with an even higher 49.3 % of the 300 who attend once a month, but only 22.3 % among those who attend more than once a week.

In addition to these nine true-false questions, the GSS included two astronomical ones phrased in a different way but integrated with those nine. The first asked, "Now, does the Earth go around the Sun, or does the Sun go around the Earth?" I won't quibble by saying they both go around their mutual center of gravity, or both orbit the center of the galaxy on complex paths determined by many gravitational influences, but grant the Earth goes around the Sun. Of 1,972 who answered this question plus the one on space funding, 80.3 % got it right. Among them, 18.2 % felt too little was being spent on space exploration, compared with a tiny 5.7 % of those who got it wrong, for a gamma of 0.33.

The second astronomical question was asked only of respondents who answered the first one correctly: "How long does it take for the Earth to go around the Sun: 1 day, 1 month, or 1 year?" Of 1,449, 78.8 % correctly answered "1 year," and 20.5 % of them wanted space funding increased. Only 30 selected the "1 month" answer, so it may not be indicative of much that 4 of these confused people, 13.3 %, wanted funding increased. Among the 279 who incorrectly answered "1 day," 12.2 felt funding was currently too little.

Two other questions superficially focused on genetics but more fundamentally assessed the respondent's ability to analyze probabilities. Both followed from this introduction: "A doctor tells a couple that their genetic makeup means that they've got one in four chances of having a child with an inherited illness." Then the interviewer asked, "Does this mean that if their first child has the illness, the next three will not have the illness?" Probability theory begins with the assumption that cases are independent, so the correct answer was "no." Of 1,740 who answered and expressed a view on space funding, 89.2 % answered correctly, and 15.9 % of them wanted space funding increased. Those who answered incorrectly included only 9.5 % feeling the funding was too little, for a gamma of 0.24 between answering correctly and wanting funding increased.

The other question in this probability pair asked, "Does this mean that each of the couple's children will have the same risk of suffering from the illness?" Of 1,958 respondents, 78.1 % correctly answered "yes," and 15.6 of them wanted space funding increased, compared with 13.3 % of those giving a wrong answer, for a gamma of 0.11. Actually, the premise of these questions seems poorly phrased, assuming that the respondent is conversant not so much with

Table 6.3 Conceptions of the scientific method

Statements about science	Very important (%)	Too little (%)	Gamma
The researchers carefully examine different interpretations of the results, even ones they disagree with	73.3	16.5	0.19
Other scientists repeat the experiment, and find similar results	67.5	15.6	0.09
It is done by scientists employed in a university setting	32.0	15.6	0.01
The conclusions are based on solid evidence	82.1	15.3	0.09
The results of the research are consistent with common sense	41.7	15.3	−0.08
The research takes place in a laboratory	40.2	14.7	−0.06
The people who do it have advanced degrees in their field	64.9	14.3	0.00
The results of the research are consistent with religious beliefs	11.2	5.2	−0.22

mathematical probability theory as with colloquial ways of expressing it, and "one in four chances of having a child with an inherited illness" might better have been expressed, "one in four chances that any given child they have will inherit the propensity for an illness." The original wording could be interpreted to mean that they had one in four chances of ever have such a child, regardless of how many others they had, thus making the "wrong" answer to the first question the correct one.

An interesting sub-module of questions about the practice of science was included only in the 2006 GSS, thus giving us fewer respondents to work with, from 858 to 870 depending on the item. While handing the respondent a card give the desired response scale, the GSS interviewer would say, "Now I'm going to read you some statements about science and scientists. Please look at Card B19. How important are each of the following in making something scientific?" Card B19 listed these four possible responses: Very important, Pretty important, Not too important, and Not important at all. Table 6.3 lists the items from highest to lowest percentage among those answering "very important" who felt that funding for the space program was too little.

The first and last items in the table have a moderate negative correlation with each other, measured by a gamma −0.22 based on 1,751 people who responded to both. Clearly, we need to consider the possible impact of religious views on the meaning of science for a significant fraction of the general public.

6.4 The Influence of Religion

Over the years, 7,865 GSS respondents expressed an opinion on space funding and answered a question concerning their beliefs about God. The majority, 4,873, selected this answer: "I know God really exists and I have no doubts about it."

Of them 11.7 % said space funding was too little. A much smaller group of 228 Atheists said, “I don’t believe in God.” A somewhat larger minority consisting of 362 Agnostics selected this answer: “I don’t know whether there is a God and I don’t believe there is any way to find out.” Among Atheists, 22.4 % felt current space funding was too little, and 24.9 % of Agnostics agreed with them.

Few Americans are willing to brand themselves as Atheists or Agnostics, yet many are not really religious. Among the most often asked questions in the GSS was “How often do you attend religious services?” Earlier we examined this item using data only from recent surveys. Of a grand total of 30,981 answering that question and the space funding one, fully 4,927 or 15.9 % said they never attend religious services. At the opposite end of the attendance scale, 2,347 or 7.6 % said they attended more than once every week. Among these high-attenders, only 7.0 % felt space funding was too little, compared with 57.5 % who felt too much was being spent. Among those who never attend religious services, 16.9 % felt space funding was too little, and 43.6 % felt it was too much. All the many other rates of attendance varied rather smoothly between these extremes. Among the 6,256 who attend every week, 9.3 % felt space funding was too little, and 48.4 %, too much. The comparable figures for the 2,256 attending once a month were 13.4 % and 45.7 %.

In 1988, the GSS included a topical module on religion, that included four items measuring attitudes toward science that might correlate with the respondent’s degree of religious traditionalism. Here are the four statements with the percentages among those who agreed saying that space funding was too little, for between 657 and 660 respondents total, and 150–267 holding these opinions:

22.7 % Science will solve our social problems like crime and mental illness.
12.0 % One trouble with science is that it makes our way of life change too fast.
10.7 % One of the bad effects of science is that it breaks down people’s ideas of right and wrong.
9.3 % Scientists always seem to be prying into things that they really ought to stay out of.

These four statements stress the moral meaning of science, rather than technical meanings. When sociologist Emile Durkheim argued a century ago that God was a metaphor for society, he was suggesting that religious faith was an expression of social solidarity and a source of moral guidance, rather than a statement about facts of nature that might put it into conflict with science (Durkheim 1915). American psychologist William James put forward a distinctive yet compatible view that it was entirely appropriate for an individual to have faith in his society’s religion, even in the absence of empirical evidence for its objective truth (James 1896). Indeed, James argued that the truth of a belief should be evaluated solely in terms of whether it was pragmatically beneficial for the believer (James 1948). This suggest that there are two alternative theories of truth, parallel to those we identified earlier for science:

Normative: A statement is true to the extent that it provides guidance for human action that leads to beneficial results.
Descriptive: A statement is true if it models the external world more accurately than alternative statements on the same topic.

Professional scientists would like to believe that scientific truth satisfies both definitions, but this assertion has not itself been verified scientifically. It is entirely possible that some religious beliefs serve human well-being, whether by supporting moral behavior or by providing psychological comfort for believers, yet contradict the findings of science. The relevance to spaceflight is precisely the tension between Heaven and the heavens. Are the skies above our heads our ultimate dwelling place, or a meaningless vacuum?

A question asked in other GSS administrations, as well as 1988, determined what factions of 3,720 respondents agreed or disagreed with this statement: "Those who violate God's rules must be punished." Among those 620 who agreed strongly, just 10.8 % felt space funding was too little, compared with 14.5 % of those 1,138 who agreed somewhat, 16.5 % of those 1,142 who disagreed somewhat, and 21.2 % of those 820 who disagreed strongly.

The number of respondents in the 1988 GSS religion module who answered the space funding items are somewhat small, but several suggestive results are worth reporting. Of 668 respondents, 69.5 % answered "yes" in response to: "When you were growing up, did anyone usually say grace or give thanks to God aloud before meals at home?" Among those from homes where thanks were given to God, 16.2 % felt space funding was too little, compared with 25.0 % from homes that did not practice this religious ritual. Of 424 respondents who are members of a church or synagogue, 13.7 % feel space funding is too little, compared with 27.7 % of the 249 who are not members.

A much larger number of respondents answered this question over the years: "Which of these statements comes closest to describing your feelings about the Bible? 1. The Bible is the actual word of God and is to be taken literally, word for word. 2. The Bible is the inspired word of God but not everything in it should be taken literally, word for word. 3. The Bible is an ancient book of fables, legends, history, and moral precepts recorded by men." Of 4,434 who considered the Bible to be the literal word of God, only 8.6 % felt too little was being invested in space exploration, compared with 14.5 % of the 6,561 for whom the Good Book was only inspired by God. A smaller secularized group of 2,248 said it was a book of fables, and 22.8 % of them felt too little was spent on space.

Another item frequently included in the GSS concerned the right of an opponent of religion to speak out against it: "There are always some people whose ideas are considered bad or dangerous by other people. For instance, somebody who is against churches and religion. If such a person wanted to make a speech in your (city/town/community) against churches and religion, should he be allowed to speak, or not?" Of 13,958 respondents who over the years felt this opponent of religion should be allowed to speak, 15.3 % felt space funding was too little, compared with only 6.1 % of the 5,874 who wanted to prohibit anti-religious speech.

The percent saying too little is being invested in space varies significantly across the five largest denominational categories having at least 500 members in the GSS: Protestant (10.8 %), Catholic (12.4 %), Jewish (16.0 %), Other (21.0 %), and None (22.5 %). Of 3,543 who pray several times a day, the fraction saying too little is 10.9 %, compared with fully 26.2 % among the 711 who never pray.

A series of questions asked in 1983 and 1984 inquired, "When you think about God, how likely are each of these images to come to your mind?" For example, the concept of God as redeemer revealed big differences. Among the 892 respondents who said "redeemer" was extremely likely to come to mind, 12.3 % thought space funding was too little, compared with 13.4 % of the 395 answering "somewhat likely," 26.7 % of the 101 who said "not too likely" and fully 35.7 % of the mere 70 who said "not likely at all."

The relationship between religion and science is complex, and some forms of faith seem to promote science rather than conflict with it (Hefner 1997; Westfall 1958). The natural theology movement of the nineteenth century promoted science as a way of learning more about God through study of his creations (Gillispie 1951). Among the theories in this area that sociologists have debated are that monotheism in general, or Protestantism more specifically, may have been a necessary precondition for the modern rise of science. Monotheism assumes that all of nature is governed by one set of laws, that are probably rational, while polytheism is more compatible with a chaotic universe in which natural laws conflict with each other. Protestantism placed unusual responsibility on each individual to develop a religious orientation, and perhaps what Max Weber called its *worldly asceticism* also promoted the rationality of science as well as investment capitalism (Weber 1930). This then raises the question of what impact recent untraditional religious and pseudoscientific movements may be having.

6.5 Pseudoscience

Supporters of the space program seldom discuss its possible connections with pseudoscience and the occult, perhaps because they are disreputable, yet they are worth considering. The most significant of the early spaceflight pioneers, Hermann Oberth, was also a believer that flying saucers were real spaceships from other worlds, and a proponent of radical occult ideas, for example in his 1959 book, *Stoff und Leben* (Oberth 1959). Here is how I presented this observation in my 1976 book on the spaceflight social movement:

> Oberth's biographers fail to mention that his rocket work was conducted simultaneously with the development of a theosophical system that must be described, delicately, as variant, if not deviant. He invented a number of vitalistic notions, including a doctrine that each cell in the body has its own immortal soul. This led him to a belief in reincarnation. In a sense, when Oberth worked toward an interplanetary future culture, he was preparing a world to dwell in after his death, reincarnated perhaps as a spaceship captain of the distant future! In 1930 Oberth published a pamphlet containing his occult ideas, incorporating the theories he had developed over the previous eight years. A book was ready for publication in 1938, but a series of difficulties prevented publication until 1959. In letters to Max Valier, Oberth urged that publications on rockets and spaceflight should not contain spiritualistic ideas so that the technical content would not be misunderstood or discredited. This was a wise strategy (Bainbridge 1976b).

This strategy was wise not only for Oberth and Valier in the 1920s, but for enthusiasts of spaceflight today. There are many motives that might inspire support, and many ways of conceptualizing interplanetary flight. But the very fact that major governments of the world have poured vast sums into rocket development may have obscured the possibly disreputable beliefs that inspired some of the pioneers, and some of today's supporters among the general public. It is not easy to evaluate the extent to which the spaceflight social movement depends ultimately upon non-empirical, even supernatural hopes, but the oldest book in this tradition of which I am aware was originally published in 1758 by cult leader Emanuel Swedenborg, who claimed that angels on the other planets were speaking with him telepathically about their worlds (Emanuel Swedenborg 1839).

The pseudoscience that might seem most closely associated with spaceflight is astrology, since both are concerned with the Moon and planets. The 2006–2010 General Social Survey topical module on science asked, "Do you ever read a horoscope or your personal astrology report?" Of 2,076 GSS respondents, 1,141 said "yes" and 14.6 % of them felt space funding was too little. This was not significantly different from the 14.8 % of the 935 who never consulted their horoscope. A related question asked: "Would you say that astrology is very scientific, sort of scientific, or not at all scientific?" Of a mere 107 who said astrology was very scientific, 11.2 % felt space funding was too little, compared with 13.0 % of the 583 who called it sort of scientific, and 16.1 % of those considering it not at all scientific.

Logically, these two astrology questions connect, and they have a reasonably strong gamma correlation of 0.34 for 4,850 respondents. But the relationship is a bit lopsided. Among people who considered astrology very scientific, 67.0 % ever read their horoscopes, compared with an essentially identical 67.3 % among those calling it sort of scientific. But just 48.7 % of those calling it not at all scientific ever read their horoscopes. For this last group, and some of the others as well, astrological horoscopes might have been fun but untrustworthy. So, reading one's horoscope is not connected with support for space funding, but feeling astrology is scientific is negatively correlated.

However, other datasets hint at positive connections between the occult and spaceflight, at least by some measures and for some respondents. The GSS question about the proper funding for the space program concerns the sober topic of appropriation of tax revenues, whereas other questions such as those about flying to other planets may tap much more mystical cultural roots. As noted in the previous chapter, the Spring 1998 Southern Focus Poll conducted by the University of North Carolina at Chapel Hill asked, "Would you favor or oppose sending a manned rocket to land on Mars?" Another question asked, "Do you believe in astrology, that is, that the position of the stars and planets can affect people's lives?" Of 276 respondents who believed in astrology, 67.4 % favored a manned landing on Mars, compared with 64.4 % of the 466 who did not believe. This is a small difference, and replicates the finding from the similar question in the GSS that astrology is not a major factor encouraging enthusiasm for the space program.

However, three other superstition variables show significant connections to Mars. One question asked, "Do you believe in ESP, that is, extra-sensory perception?" Of the 495 respondents who believe in ESP, 72.1 % favor a Mars landing, compared with only 54.7 % of the 236 who do not believe. Another pseudoscience item was: "In your opinion, are UFOs something real, or just people's imagination?" Of the 309 people who believed UFOs are real, 74.1 % Favored a Mars landing, compared with 56.1 % of the 300 who considered UFOs to be imaginary. The third item was: "Do you believe that extraterrestrial beings have visited Earth at some time in the past?" Of the 274 people who believed in ancient astronauts, 75.5 % favored the Mars landing, versus 56.8 % of those 387 who did not. The ESP item had nothing substantively to do with spaceflight and showed the same pattern as the UFO and extraterrestrials items, which suggests we are dealing with a general pseudoscientific enthusiasm, rather than one focused on the planets.

In the late-1970s I carried out a pair of modest questionnaire studies with students at the University of Washington, but with a novel preparation (Bainbridge 1978). On one day, the class saw a pseudoscientific documentary film, *In Search of Ancient Astronauts*, that favorably presented the crank theory of Erich von Däniken that ancient artworks depict extraterrestrial visitors to Earth. At the next class meeting, students completed a questionnaire, then received a lecture debunking the film. One of the classes also heard a brief lecture on the pseudoscientific biorhythm theory, that mathematically precise rhythms shape human behavior in three multi-day sine waves. Opinions of the ancient astronauts theory were measured by two oppositely phased items: "von Däniken's theory of ancient astronauts is probably true," "von Däniken's theory of ancient astronauts is probably false." Correlations between these items and others were almost identical, except with opposite signs, indicating that yea-saying basis was not a problem for this topic. One questionnaire included a brief astronomy quiz, plus items about the science courses a student had taken, but these did not at all predict attitudes toward ancient astronauts. Here are Pearson's *r* correlations between accepting von Däniken's theory of ancient astronauts and rating the truth of five other theories on a scale from 0 to 100, based on responses from a class of 114 students:

0.43 There is intelligent life on other planets.
0.38 Biorhythm theory is true.
0.52 Extrasensory perception exists.
0.01 Miracles actually happened just as the Bible says they did.
0.01 Darwin's theory of evolution is true.

It is logical that believers in von Däniken's theory also think extraterrestrials exist, but there is no logical connection to biorhythms and ESP, but we see three strong positive correlations. We see no correlation with a common article of religious faith, or acceptance of a standard scientific theory. Apparently, von Däniken's theory is part of a generalized but distinctive pseudoscientific subculture that is as remote from conventional religion as it is from real science. Another class of 121 students produced a tau correlation of −0.42 between the von Däniken item and "UFOs are probably illusions." Here are five other correlations:

0.54 UFOs are probably spaceships from other worlds.
0.30 I myself have had an experience which I thought might be an example of extrasensory perception.

0.39 Extrasensory perception probably exists.
0.36 Some Eastern practices, such as Yoga Zen, or Transcendental Meditation, are probably of great value.
0.40 There is much truth in astrology.

The social organization of pseudoscience and the occult is far more chaotic than is traditional religion, yet both are transmitted through social relationships and a diversity of communication channels. Science is taught in schools, but after graduation most people must rely upon the mass media for their instruction.

6.6 The GSS News Interest Module

For most adult respondents to the General Social Survey, education and other factors like social class origins influenced attitudes earlier in their life histories, but today's attitudes also reflect current information inputs. One item in the 2006–2010 module compared the channels through which information flows: "We are interested in how people get information about science and technology. Where do you get most of your information about science and technology (newspapers, magazines, the Internet, books or other printed materials, TV, radio, government agencies, family, friends, colleagues, or some other source)?" Of 2,119 people who both answered this question and expressed a view on space program funding, by far the two most popular information media were TV with 840 respondents and Internet with 594. Support for space investment differed greatly between these two groups, only 10.8 of the TV group feeling that funding was too little, compared with fully 20.7 % of the Internet users.

Over three surveys in 2008–2012, the GSS asked respondents how much they were interested in news about space exploration, in a battery of ten items sharing this introduction: "There are a lot of issues in the news, and it is hard to keep up with every area. I'm going to read you a short list of issues, and for each one—as I read it—I would like you to tell me if you are very interested, moderately interested, or not at all interested?" The focus on the news made responses somewhat dependent upon what actually had been in the news recently, but inclusion in a list of news topics makes it possible to compare public space consciousness with competing interests, several of which relate to science and technology.

Interest in news concerning "issues about space exploration," correlates very strongly with the original one, natspac which Chap. 3 used to great effect, which asked about the appropriate level of funding for the space program. Among 374 respondents who were not at all interested in space exploration, only 3.2 % felt too little was being invested, in comparison with 67.1 % who felt too much was. The 159 people who were moderately interested included 9.7 % who wanted funding increased, compared with 30.2 % who wanted it decreased. There was a huge difference between the two groups in the fraction who felt funding was about right, 60.2 % of the moderately interested, compared with 29.7 % of those not at all interested.

The figures for those 270 respondents who were very interested in space were striking, if perhaps predictable. Of these enthusiasts, 40.4 % felt too little was being spent, 48.1 % felt funding was about right, and only 11.5 % felt too much was being invested. Clearly these two items are very strongly correlated, undoubtedly because they share much of the same meaning. Arguably the right correlation coefficient to express the relationships in a table of 3×3 cells, without strict numerical scales, is tau-b, which in this case is 0.46, quite large for that coefficient. Perhaps the second-best choice on mathematical grounds would be gamma, which tends to be larger in data such as these, and is 0.68. When we do factor analysis, we begin with Pearson's r correlations, which may not strictly be correct in terms of statistical theory but can be a useful tool on practical grounds, and r here is 0.50. All these are huge in the context of questionnaire data. Table 6.4 examines all ten items in the news interest module, in a way that facilitates comparison between interest in space exploration and the other nine items.

The news topics are arranged in terms of declining correlation with interest in space exploration, and of course the first row reminds us that the correlation of one item with itself is 1.00. The number of respondents reflected in column 3 ranges from 3,176 to 3,187, while the range for column 4 which adds the space funding item is from 1,168 to 1,172. The essentially zero correlations in the bottom row, for "local school issues," is yet another example in which spaceflight seems unconnected to education, despite all the rhetoric that it is.

One obvious reason why the correlations are higher in column 3 than column 4 is that the ten items all measure general interest in news reports, as well as interest in a specific topic within the news. Also, there may be a social desirability factor at work, in that people may feel they are expected to be interested in the news, and some of them will therefore express interest that they actually lack, across all ten items. Differences comparing one row of the table with another may be more valid. Part of the column differences may be real, of course, because people may give the space program a low priority for government funding, despite being

Table 6.4 Interest in news topics and spaceflight enthusiasm

Topic in the news	Percent very interested (%)	Tau-b correlation	
		Space interest	Space funding
Issues about space exploration	21.8	1.00	0.46
Issues about new scientific discoveries	40.5	0.46	0.23
Issues about the use of new inventions and technologies	41.3	0.42	0.19
Issues about environmental pollution	48.1	0.30	0.11
International and foreign policy issues	23.4	0.30	0.16
Issues about new medical discoveries	60.3	0.29	0.08
Issues about military and defense policy	39.8	0.25	0.17
Economic issues and business conditions	49.4	0.21	0.12
Agricultural and farm issues	24.7	0.19	0.04
Local school issues	51.3	0.06	0.03

interested in its results, because the money could otherwise be spent on something they felt had more tangible benefits, or indeed returned to the taxpayer.

6.7 Science and Superstition

The relationship between science and the space program, in the public mind, is complicated by public ambivalence about science more generally. Scientists are highly admired, yet the public does not consult them about central questions like the existence of God, the implications of biological evolution for the modern family, or the causes and solutions for income inequality and congressional gridlock. Given this situation, government science agencies sell their value in terms of hard-technology engineering advances, rather than intellectual progress.

If many scientists seek the truth so that humans will no longer be misled, societal leaders in politics, religion, finance, industry, and the mass media may feel their own authority is threatened. Yes, a statement is true if it models the external world more accurately than alternative statements on the same topic. But a statement is useful to particular people if it provides guidance for action that leads to beneficial results for themselves, including action that spreads false rumors. The public apparently does not want scientists to disabuse them of their illusions, and indeed some of those illusions may be favorable for spaceflight, even as others work against it.

Yet from an evolutionary standpoint, this may not be entirely bad. The occult illusions that burdened Hermann Oberth did not prevent him from completing very accurate and detailed technological assessment of how liquid-fuel rockets could—just barely—be used to send small payloads into space. But the personality characteristics that interested him in the occult may also have given him unrealistic optimism about the future applications of these low-energy rockets. The net result was that he accomplished worthwhile advances, quite apart from any support from public opinion, the scientific community, or private industry. Anything that liberates the imagination can cause many people to fail in unrealistic endeavors, but cause a few to succeed spectacularly in directions that conventional people would never have gone. That is to say, space travel is a fact that benefits from episodes of fiction.

References

Bainbridge, W. S. (1976a). *The spaceflight revolution*. New York: Wiley-Interscience, p. 7.

Bainbridge, W. S. (1976b). *The spaceflight revolution* (p. 31). New York: Wiley-Interscience.

Bainbridge, W. S. (1978). Chariots of the gullible. *Skeptical Inquirer, 3*(2), 33–48.

Bainbridge, W. S. (1998). Extraterrestrial tales. *Science, 279*(5351), 671.

Bainbridge, W. S. (2003). Privacy and property on the net: Research questions. *Science, 302*(5651), 1686–1687.

Bainbridge, W. S. (2007). The scientific research potential of virtual worlds. *Science, 317*(5837), 472–476.

Durkheim, E. (1915). *Elementary forms of the religious life*. London: G. Allen and Unwin.
Emanuel Swedenborg, E. (1839). *Concerning the Earths in our solar system*. Boston: O. Clapp.
Etzioni, A. (1964). *The moon-doggle: Domestic and international implications of the space race*. Garden City, New York: Doubleday.
Etzioni, A. (1966). On the national guidance of science. *Administrative Science Quarterly, 10*(4), 466–487.
Gillispie, C. C. (1951). *Genesis and geology: A study in the relations of scientific thought, natural theology, and social opinion in Great Britain, 1790–1850*. Cambridge, Massachusetts: Harvard University Press.
Guth, A. H. (1981). Inflationary universe: A possible solution to the horizon and flatness problems. *Physical Review D, 23*(2), 347–356.
Hefner, P. (1997). The science-religion relation: Controversy, convergence, and search for meaning. *International Journal for the Psychology of Religion, 7*(3), 143–157.
James, W. (1896). The will to believe. An address to the philosophical clubs of Yale and Brown universities, published in *New World*, June. falcon.jmu.edu/~omearawm/ph101willtobelieve.html.
James, W. (1948). *Essays in pragmatism*. New York: Hafner.
Launius, R. (2006). Assessing the legacy of the Space Shuttle. *Space Policy, 22*, 226–234.
Lewis, C. T. (1916). *A Latin dictionary for schools* (p. 942). New York: American Book Company.
NASA Historical Staff, Office of Policy Planning, Scientific and Technical Information Division. (1964). *Astronautics and Aeronautics, 1964: Chronology on Science, Technology and Policy*. Washington, DC: NASA, p. 262.
National Science Board. (2012). *Science and engineering indicators*. Arlington, Virginia: National Science Foundation.
Oberth, H. (1959). *Stoff und Leben*. Remagen, Germany: Reichl.
van Allen, J. A. (2004). Is human spaceflight obsolete? *Issues in Science and Technology*. Retrieved December 19, 2013, from http://www.issues.org/20.4/p_van_allen.html.
Weber, M. (1930). *The Protestant ethic and the spirit of capitalism*. New York: Scribner.
Westfall, R. S. (1958). *Science and religion in seventeenth-century England*. New Haven, Connecticut: Yale University Press.
Whithey, S. B. (1959). Public opinion about science and scientists. *The Public Opinion Quarterly, 23*(3), 382–388.

Chapter 7
Literature

Within the subculture of "SF fans," a standard proverb, attributed either to Arthur C. Clarke or Isaac Asimov, is "Science fiction is escape into reality." Legend has it that these two leading writers once shared a taxicab and argued which of them was the greatest writer. After heated debate, they agreed that one was the greatest writer of science fiction, while the other was the greatest writer of science fact. Unfortunately, however, they could not agree which was which. This raises the issue of the extent to which all enthusiasm for spaceflight, fictional or factual, is a form of escapism for people who are uncomfortable living on Earth. Jack Lait and Lee Mortimer defined science fiction as "a genre of escape literature which takes the reader to far-away planets—and usually neglects to bring him back (Lait and Mortimer 1953)."

The article about spaceflight in the online *Encyclopedia of Science Fiction* says: "It is natural that sf should be symbolized by the theme of space flight, in that it is primarily concerned with transcending imaginative boundaries, with breaking free of the gravitational force which holds consciousness to a traditional core of belief and expectancy." The article then goes on to demonstrate that almost all depictions of space travel in SF violate known principles of science and engineering.

Scientists have every reason to be ambivalent about science fiction. On the one hand, it inspires readers to value science, and on the other it spreads many unscientific myths (Milburn 2010). Robert A. Heinlein, often called "the dean of science fiction writers," preferred to call the genre *speculative fiction* (Heinlein 1964, 1969). Considered as philosophical speculation, "SF" need not misinform the reader so much as expand the reader's perspective, and exercise the reader's mind. To the extent that some of its speculations can be framed as hypotheses, then they can be subjected to empirical verification. To the extent that others set goals, then perhaps technologies can be devised to accomplish them. Spaceflight may be the best example, because it has played so many roles in so many quite different stories, that it offers a spectrum of valuable concepts.

W.S. Bainbridge, *The Meaning and Value of Spaceflight*,
Space and Society, DOI 10.1007/978-3-319-07878-6_7

7.1 Early Flights of Fantasy

It is hard to say when the modern idea of spaceflight emerged, and within literature and the arts there always existed alternative conceptions of the starry universe. In the thirteenth century poems of the *Carmina Burana*, made famous in the twentieth century by composer Carl Orff, a prominent chorus considered the Moon as pure metaphor: "O Fortuna velut luna statu variabilis..." (O Fortune, like the Moon you are changeable...) (Patch 1922). Fortuna is not the statistical concept of random probability, but a goddess. Gustav Holst's beloved 1916 orchestral suite, *The Planets*, is astrological not astronomical. There is room to debate whether Cyrano de Bergerac's seventeenth-century satirical novel about a lunar flight should be admired as a modern conception, because the Moon is depicted as a world rather than a goddess, or snickered at because it imagined one could fly there by capturing dew in bottles, because dew lifts from the grass heavenward after the sun rises (de Bergerac 1899). Jules Verne's widely-read 1865 novel about the first lunar flight is more realistic, using a gigantic gun to fire a manned space capsule, but it glosses over the fact that the acceleration at launch would have been fatal (Verne 1869).

In *Auf Zwei Planeten* (*On Two Planets*), an 1897 spaceflight novel by the German philosopher Kurd Lasswitz, a balloon expedition to the North Pole discovered the Martians had established a base there, initiating a complex interaction between two worlds that raised issues such as the theory that a technologically advanced species would also be morally advanced (Lasswitz 1969). The year before, Lasswitz had published a philosophical essay, "Nature and the Individual Mind," that argued against dualistic theories that consider mind and matter to be separate forms of reality. Throughout, he returns constantly to the image of a human being contemplating the Moon:

> If we call the moon a body which is defined by its position, size, weight, and motion, and which existed long before men or human consciousness existed, then, of course, the moon is declared to exist as an independent regular system, quite independently of the existence of man. But the determinative data of space, time, size, weight, etc., which represent the thing moon, are exactly the same as the data which we find now in consciousness, since men and astronomers exist. The form of consciousness we can neither take from them nor give to them. It follows, therefore, that these laws which now we have discovered to be laws of *human* consciousness, are naught else than the *universal* laws under which the development of nature proceeds and ever has proceeded, and under which we human beings, too, have been developed. In other worlds, the laws under which alone nature can be represented and our own present existence in nature understood, may just as well be termed laws of nature as laws of consciousness. The two are identical. (Lasswitz 1896)

This passage, written well over a century ago, establishes two related principles that define not only science fiction but the general modern consciousness. First, that humans and astronomical objects are elements within the same natural system, connected and capable of interacting meaningfully with each other. Second, as the use twice of the word *data* suggests, existence can be conceptualized as a coherent system of information, dynamic but following logical laws.

Seldom read, and hardly ever praised, an 1898 spaceflight novel by American astronomer Garrett P. Serviss actually deserves serious consideration. Titled *Edison's Conquest of Mar*s, it was commissioned by a Boston newspaper that had published an altered version of *War of the Worlds* by H. G. Wells, shifting the setting from England to the United States (Wells 1898). Serviss began where Wells left off, imagining that Thomas Edison developed spaceship technology that improved upon the Martian variety, as well as new weapons. An international campaign collected wealth from many nations to build a fleet of a hundred warships, not merely for revenge but in awareness that the Martians might attack again if they figured out how Earth's bacteria had defeated them when mere humans could not. This sounds like a rather crude rip-off novel, but Serviss used the opportunity to give readers a series of reasonably authoritative science lessons about the solar system. His greatest challenge, as a pioneer American science fiction author, was how to conjure up a plausible means of propulsion for the spaceships:

> It would carry me into technical details that would hardly interest the reader to describe the mechanism of Mr. Edison's flying machine. Let it suffice to say that it depended upon the principal of electrical attraction and repulsion. By means of a most ingenious and complicated construction he had mastered the problem of how to produce, in a limited space, electricity of any desired potential and of any polarity, and that without danger to the experimenter or to the material experimented upon. It is gravitation, as everybody knows, that makes man a prisoner on the earth. If he could overcome, or neutralize, gravitation he could float away a free creature of interstellar space. Mr. Edison in his invention had pitted electricity against gravitation. Nature, in fact, had done the same thing long before. Every astronomer knew it, but none had been able to imitate or to reproduce this miracle of nature. When a comet approaches the sun, the orbit in which it travels indicates that it is moving under the impulse of the sun's gravitation. It is in reality falling in a great parabolic or elliptical curve through space. But, while a comet approaches the sun it begins to display - stretching out for millions, and sometimes hundreds of millions of miles on the side away from the sun - an immense luminous train called its tail. This train extends back into that part of space from which the comet is moving. Thus the sun at one and the same time is drawing the comet toward itself and driving off from the comet in an opposite direction minute particles or atoms which, instead of obeying the gravitational force, are plainly compelled to disobey it. That this energy, which the sun exercises against its own gravitation, is electrical in its nature, hardly anybody will doubt. The head of the comet being comparatively heavy and massive, falls on toward the sun, despite the electrical repulsion. But the atoms which form the tail, being almost without weight, yield to the electrical rather than to the gravitational influence, and so fly away from the sun. (Serviss 2009)

This passage reveals one of the chief virtues of *Edison's Conquest of Mars*, in communicating much real science to the general reader in the context of speculation about future possibilities for spaceflight. Of course, a contemporary astronomer would describe the solar wind rather than electrical repulsion, in explaining a comet's tail, but such is progress. By casting a real living person as his hero, Serviss established another link between fiction and reality. The novel also conveys a strong sense that achievement of spaceflight could express supreme human values, in at least three ways. First, the heroes refuse to be discouraged, even by the Martians' complete destruction of New York City, investing powerful will and intellect in rising up from adversity. Second, nationalistic sentiments in favor of American exceptionalism are depicted as idealistic achievement rather than arrogance. Third, the people of the Earth unify to defeat a common enemy, rising above their petty conflicts by means of spaceflight. Among those explicitly inspired by Serviss and the bowdlerized newspaper version of Wells was Robert H. Goddard, the real-world American spaceflight pioneer.

Serviss also deserves credit, apparently, for being the first person to suggest publically that spaceflight would employ nuclear power, in his 1909 novel, *A Columbus of Space* (Serviss 1911). However, he did not imagine nuclear rockets, postulating instead that nuclear energy could power electrostatic propulsion as described above. When H. G. Wells sent fictional men to the Moon in 1901, they were wafted there by cavorite, a fictional substance that negates gravity (Wells 1901). Edgar Rice Burroughs employed a psychic leap comparable to reincarnation to get John Carter to Mars in 1912 (Burroughs 1917). When the science fiction subculture consolidated in the United States in the late 1920s, rockets were not yet the preferred means for spaceflight, as illustrated by the fact that E. E. "Doc" Smith's influential novel *The Skylark of Space* used an anti-gravity drive analogous to the ideas of Serviss and Wells, when it was published in the first science fiction magazine in 1928 (Smith 1946).

7.2 The Science Fiction Subculture

While many individual authors wrote about spaceflight around the beginning of the twentieth century, a science fiction subculture emerged and became culturally influential in the middle decades. Its launch was the first magazine entirely devoted to SF, *Amazing Stories*, in April 1926, edited by Hugo Gernsback, after whom the annual SF Hugo awards are named. While spaceflight did feature prominently in many stories, Gernsback's own passions were in the burgeoning electronics industry, a foreshadowing of the contemporary situation, in which information technology based on advanced electronics has much greater impact on humanity than does spaceflight, and which offers much greater opportunities for ordinary people to contribute to progress.

Historians speak of a Golden Age of Science Fiction, agreeing that it began in 1938 when John W. Campbell, Jr., became editor of *Amazing's* rival, *Astounding Stories*, which is still published today as *Analog*. There is some disagreement when the Golden Age ended, and Wikipedia claims it did in 1946. Having read every issue of *Astounding* over that period, I would prefer to place the end in 1953, marking a turning point when *Mission of Gravity* by Hal Clement was serialized in the magazine, and *Demolished Man* by Alfred Bester won the very first Hugo award for best novel but had been serialized in the upstart rival to *Astounding*, *Galaxy Science Fiction* (Clement 1954; Bester 1953). Having interviewed both authors, and read their works repeatedly over the years, I believe these two landmark novels define the territory of high quality science fiction.

Hal Clement (Harry Clement Stubbs) studied astronomy at Harvard and for many years taught chemistry and astronomy at Milton Academy, a private high school in Massachusetts. He brought to his fiction writing a keen analytical mind that sought to derive interesting stories from real principles of the very sciences he taught. The focus of *Mission of Gravity* is less the people in the story than the planet Mesklin, on which the action takes place, in orbit around the nearby star 61 Cygni A, which astronomers at the time thought might have a planet, because of anomalies in measurements of its motion that might have resulted from the gravity of a large invisible body in a system that already had a second star.

Clement reasonably guessed that a planet in a double-star system might rotate very rapidly, because tidal forces from the second star would have affected its formation. A large planet rotating rapidly could have low-enough gravity at its equator for humans to land, and indeed be exceedingly oblate rather than spherical, but have very high gravity at its poles where the centrifugal force from rotation would have been absent. In *Mission of Gravity*, an automated research vessel has landed at a pole, but fails to take off after collecting its scientifically valuable data. The high gravity prevents humans from going to get the data, so they enlist local inhabitants, rather like huge intelligent cockroaches, who can tolerate the gravity, to do the job for them. Many episodes in the story build upon the conceivable implications of these astounding but logical premises based on the physical sciences.

The Demolished Man could hardly be more different, and it employs spaceflight only as a secondary feature of the background. It was as logical as *Mission of Gravity*, but based on theories from the psychological sciences, rather than physical sciences, thus focusing on humans rather than the natural environment. A man commits a murder on a future Earth, where telepathy exists and police detectives can read minds, yet he is able to deceive the authorities about his villainous act. The logical explanation is that he has also deceived himself, at least about his motive if not the nature of his vile deed itself. Exceedingly logical, but also innovative in some aspects of literary style, *The Demolished Man* is based on two "sciences," neither of which is astronomy: parapsychology and psychoanalysis.

A decade after both novels were published, as the race to the Moon was beginning, psychoanalysis probably hit its peak popularity, but today it is almost entirely absent from academic psychology, surviving among the humanities and various corners of popular culture (Bainbridge 2012). The achievement of real spaceflight may have dampened enthusiasm for fictional astronautics, and a peak of SF magazine publishing around 1953 was followed by a manifest decline. In 1960, SF fan Earl Kemp sent a brief questionnaire to many leaders of the field, asking primarily "Who killed science fiction." Part of the response from Isaac Asimov blamed the real space program, and the inability of many authors to find a way of surpassing it:

> Obviously, Sputnik and all that followed bears a major share of responsibility for the decline of magazine science fiction. The newspapers are so science fictiony now that there is scarcely any urge to continue looking for more science fiction in the magazines. In fact, the type of science fiction in the newspapers now - all this talk of space and satellites and moonshots - is so interbound with cold war and national prestige and military brass-hats

> that it makes science fiction unsavory, even to me, for instance. This isn't helped by the fact that too large a proportion of science fiction stories written today are "tomorrow fiction" - that is, they deal with a situation one step ahead of the headlines so that one gets the impression that all science fiction is but a kind of "home life at Cape Canaveral" or "Look, Ma, the general is jutting his jaw and says the next satellite will go through carrying a man, Ma." Hell, for a group of people who have been bouncing around the galaxy for 20 years, it's downright sickening to go back to trying to reach the moon with chemical rockets.

By my own count, 403 issues of science fiction magazines were published in 1946–1950, 910 in 1951–1955, 767 in 1956–1960, and 482 in 1961–1965 (Bainbridge 1976a). Clearly, this was some kind of rise and fall. SF fans at the beginning of the 1950s imagined their genre was becoming very popular, but another plausible explanation I have suggested is that the apparent SF boom was an artifact of changes in the publishing industry, under pressure from the really booming medium of that period, television. Publishers saw profits diminishing in other areas and brought out new SF magazines to see if they could be profitable, and terminated them when they proved not to be (Bainbridge 1976b).

Circulation figures for particular magazines provide some insights, fluctuating from year to year, along with some major swings and secular trends. *Astounding Stories* became *Astounding Science Fiction* and reached its early peak circulation in 1952 at 105,700 copies of each issue. This dropped to 89,153 in 1957 and 77,124 in 1962. As *Analog*, it climbed back up to 99,228 in 1969 and 116, 521 in 1973 (Bainbridge 1976c). But today it is a mere shadow of its former self, selling just 27,803 copies of each issue in 2012 (Dozois 2013). Given the obvious importance of science and technology for American life, one might have thought the numbers for the most influential science fiction magazine would have been ten times what *Astounding/Analog* ever achieved.

However, it may be that the magazines played a crucial role in establishing the SF subculture, after which other media could take the genre to greater heights of popularity. Each issue of *Astounding* contained an editorial, a factual science article, a collection of letters to the editor from readers, and a book review section. Thus, it was a medium for communication between the authors and leading fans, more than it was a literature for the general public. As the magazines faded, books became more prominent. From 1953 through 2013, 283 novels are listed by Wikipedia as Hugo winners or finalists. Of the first hundred, which cover the period up through 1977, 40 were originally published in magazines. The next two decades, up through 1997, saw 101 novels nominated, just 13 of which had been published in magazines. Of the most recent 82, only 1 was reported to have been published in a magazine.

7.3 Three Dimensions of Science Fiction

In 1978, well after the Apollo Program and other real-world accomplishments had redefined spaceflight as a practical possibility rather than a fantasy, and after the 1960s had introduced greater literary sophistication into the SF field, I administered questionnaires to participants at the world science fiction convention in Phoenix, Arizona, in order to develop a cultural map of the genre at that point in its maturity (Bainbridge 1986). The main questionnaire chiefly consisted of preference items, asking respondents to rate 140 authors, 40 types of literature, and several other aspects of science fiction on a preference scale from 0 (do not like) to 6 (like very much). There were five versions of the questionnaire, presenting the authors in different random orders to guard against correlations that reflected mere proximity of authors to each other, and two of the 140 were in fact fake, to guard against frivolous responses. The most intensive analysis focused on 409 respondents who rated at least 50 real authors but skipped the fake authors. Mean preference scores were based on the full dataset of 595 respondents, all of whom were dedicated science fiction readers, and several of whom were also professional authors.

Especially fruitful was a factor analysis based on 276 respondents who rated at least 75 authors but neither fake author, mapping in four dimensions the 73 authors rated by at least 80 % of this group of experts. The fourth dimension, fittingly enough, turned out to be time, headed by H. G. Wells and Jules Verne, thus representing primarily the distant past. The three other dimensions represented very different ideological variations, and thus can help us understand that science fiction is not one, unified ideology, but has least these three dimensions. Among these 276 experts, I correlated factor scores with the types of literature in order to identify the factors unambiguously.

Factor I was strongly associated with liking "fiction based on the physical sciences" ($r = 0.45$). Factor II correlated with liking "fiction based on the social sciences" ($r = 0.39$), and Factor III correlated with liking "stories about magic" ($r = 0.53$). Thus, each factor reflects a different dimension of the wider intellectual culture. Even more strongly, the factors correlated with specialized terms employed by literary critics within the genre. Factor I correlated with "hard-science fiction" ($r = 0.54$), and Factor II correlated with "new-wave science fiction" ($r = 0.61$). Factor III correlated with the very general term "fantasy" ($r = 0.57$) but also with the specialized action-oriented exotic form of fantasy called "sword-and-sorcery" ($r = 0.60$). More complete definitions of the factors came from correlating the defining preference variables with others in the dataset, giving the correlations for 409 respondents and in parentheses the mean preference ratings for all 595 on the 0–6 scale. Here is Factor I:

Hard-Science (4.53)

0.66 Fiction based on the physical sciences (4.26)
0.51 Stories about new technology (4.60)
0.49 Factual science articles (3.92)
0.47 Stories which take current knowledge from one of the sciences and logically extrapolate what might be the next steps take in that science (4.83)
0.46 Stories in which there is a rational explanation for everything (3.57)
0.43 Factual reports on the space program and spaceflight (4.26)
0.33 Golden Age science fiction (3.72).

The meaning of hard-science science fiction could hardly be more clear. It is rational narratives based on what are commonly called the "hard sciences," and it is connected strongly with enthusiasm for the real-world space program. It is noteworthy that this is the only one of the three dimensions of science fiction favorable to or inspired by spaceflight. Clement's *Mission of Gravity* is a perfect example of hard-science SF. The connection to the Golden Age was confirmed with very different data way back in 1958 by Walter Hirsch who studied the professions of heroes in SF stories, finding that the overwhelming majority worked in the physical sciences (Hirsch 1958). Here is Factor II, going a little further down the correlation ladder to discern some of its components, and describing a style that emerged after the Golden Age, during the 1960s:

New-Wave (3.32)

0.65 Avant-garde fiction which experiments with new styles (3.14)
0.40 Fiction based on the social sciences (3.74)
0.39 Science fiction of the 1960s and 1970s (4.92)
0.38 Fiction that is critical of our society (3.65)
0.37 Fiction which deeply probes personal relationships and feelings (4.30)
0.34 British science fiction (3.79)
0.31 Feminist literature (2.65)
0.29 Stories in which the main character is sensitive and introspective (4.40)
0.27 Fiction concerned with harmful effects of scientific progress (3.19)
0.27 Poetry (3.01).

New-Wave was far less popular among participants at the 1978 world science fiction convention than Hard-Science, earning a mean preference score of only 3.32 compared with 4.53, yet is clearly a distinct sub-genre. As an explicit artistic movement, the New Wave was promoted by the British magazine, *New Worlds* when it was edited beginning in 1964 by Michael Moorcock, yet included many American authors and had many precursors who bridged the gap between the science fiction subculture and various movements in avant-garde literature. Indeed, we might retrospectively count Bester's *The Demolished Man* in this genre, as it even experiments with literary styles, using blank verse when depicting a conversation among telepaths. As part of the legacy of the 1960s, New Wave was somewhat politically radical and technophobic, and has been reflected more recently in a host of dystopian novels and movies. "New-wave science fiction" correlates 0.23 with "horror-and-weird," which in turn correlates 0.43 with "fantasy," suggesting that both dimensions of science fiction are more pessimistic or emotive than hard-science. Here is Factor III:

Fantasy (4.49)

0.66 Sword-and-sorcery (3.84)
0.65 Science-fantasy (4.71)
0.61 Stories about magic ()
0.58 Myths and legends (4.22)
0.47 Sagas and epics (3.96)

0.44 Stories set in a universe where the laws of nature are very different from those found on our world (4.49)
0.44 Tales of the supernatural (3.08)
0.43 Horror-and-weird (2.92)
0.40 Stories about barbarians (2.85)
0.35 Ghost stories (2.58)
0.35 Occult literature (2.13)
0.28 Poetry (3.01).

With a mean rating of 4.49, fantasy is quite popular within the science-fiction subculture, especially the kind called "science-fantasy" which is "set in a universe where the laws of nature are very different from those found on our world." Ghost stories and occult literature are rather unpopular, confirming that science fiction fans prefer rational but escapist narratives over ones based on conventional supernatural superstitions. The easiest way to define "sword-and-sorcery" is to report that *Star Wars* belongs to this sub-genre, an action-oriented mythos where light sabers and The Force are fanciful rather than scientific.

To conclude our consideration of this now rather historical study, we should identify which of the classic authors belongs to the three groups. Table 7.1 lists the most representative authors, along with their Pearson's r correlation with the type to which they belong among the 409 careful respondents, and their mean preference scores among all 595 respondents. Because Hard-Science connects to spaceflight and thus is of greatest interest here, the table lists all authors with correlations above 0.25 with this type of literature, but only those with correlations over 0.35 with New-Wave or Fantasy.

The final column of Table 7.1 lists the genre labels applied to each of the authors by Wikipedia, as of February 1, 2014. Note that Wikipedia uses all three of the labels here applied to the factors, although the terms "science fiction" and "fantasy" are used rather loosely. Wikipedia correctly assigns the best work of Isaac Asimov, and by extension of Hard-Science more generally, to the Golden Age of science fiction, and uses the term "Postmodernism" to refer to Philip K. Dick and thus to New-Wave. The term "high fantasy" applied to Tolkien generally refers to fantasy stories set in well-described worlds totally separate from our own, but only seldom described as other planets. While Michael Moorcock is correctly listed among the authors of fantasy literature, Wikipedia notes his leadership in the New Wave movement, where he served as the leading editor rather than as a prominent writer.

7.4 Dimensions of Space Program Support

As described in Chap. 4, questionnaire S1986A was administered to Harvard students primarily to collect their ideas about the possible goals and thus values of the space program, using a few open-ended items. But it also included fixed-choice

Table 7.1 Leading authors of the three science fiction dimensions

Author	*r* with type	Mean rating	Fields, genres, or literary movement from the author's biography in wikipedia
Hard-science			
Hal Clement	0.38	4.18	Science fiction, hard science fiction
Jack Williamson	0.38	4.13	Science fiction
Larry Niven	0.35	5.06	Hard science fiction
Arthur C. Clarke	0.35	4.93	Hard science fiction, popular science
E. E. "Doc" Smith	0.34	3.48	Science fiction (notably space opera)
Murray Leinster	0.33	4.10	Fantasy, science fiction, horror fiction, mystery fiction, Western fiction, general pulp fiction
John W. Campbell, Jr.	0.31	3.97	Science fiction
Fred Hoyle	0.30	3.37	Astronomy
Jerry Pournelle	0.29	3.96	Science fiction
Fred Pohl	0.29	4.56	Science fiction
Isaac Asimov	0.28	5.08	Science fiction (hard SF, social SF), mystery, golden age of science fiction
Poul Anderson	0.27	4.87	Science fiction, fantasy, time travel, historical fiction
Robert A. Heinlein	0.27	5.05	Science fiction, fantasy
New-wave			
Harlan Ellison	0.52	4.01	Speculative fiction, science fiction, fantasy, crime fiction, mystery, horror, film and television criticism, New Wave
Barry Malzberg	0.41	2.64	Science fiction, fantasy
Kate Wilhelm	0.40	4.19	Science fiction, mystery, fantasy
Robert Silverberg	0.39	4.52	Science fiction, fantasy
Kurt Vonnegut	0.39	3.36	Satire, gallows humor, science fiction
Samuel R. Delany	0.39	4.02	Science fiction, fantasy, autobiography, creative nonfiction, erotic literature, literary criticism, new wave
Norman Spinrad	0.38		Science fiction author, essayist, critic
Philip K. Dick	0.37	3.83	Science fiction, paranoid fiction, postmodernism
Judith Merril	0.36	3.36	Science fiction
Joanna Russ	0.35	3.02	Science fiction, fantasy, feminist literary criticism
Fantasy			
J. R. R. Tolkien	0.48	4.73	Fantasy, high fantasy, translation, criticism
A. Merritt	0.43	3.58	Speculative fiction, supernatural fiction
Fritz Leiber	0.40	4.85	Fantasy, horror, science fiction
Robert E. Howard	0.38	3.50	Sword and sorcery, westerns, boxing stories, historical, horror
H. Rider Haggard	0.38	3.47	Adventure, fantasy, fables, romance, sci-fi, historical
Michael Moorcock	0.37	3.77	Science fiction, fantasy, historical fiction, New Wave science fiction

items about the space program, plus a set of preference questions that included some derived from the 1978 science fiction convention study. Table 7.2 correlates items reflecting the three dimensions of science fiction, as defined by dedicated fans, with support for the space program. The question was similar but not identical to the space funding item: "Should the amount of money being spent on the U.S. space program be increased, kept at current levels, decreased, or ended altogether?" The table is based on 512 Harvard students who responded to most questions, and including equal numbers of males and females to control for the large effects associated with gender. Of these respondents, 195 wanted funding increased; 222 wanted it kept at current levels; 74 wanted it decreased, and 7 wanted it ended. The fraction preferring an increase, 39 %, is much higher than what we saw for the

Table 7.2 Science fiction preferences and space funding among Harvard students in 1986

Preference for:	Mean preference among those who want space program funding…			Correlation (gamma) with increased
	Increased	Kept at current levels	Decreased or ended	
Science fiction	4.17	3.25	2.44	0.38
Hard-science SF items				
Fiction based on the physical sciences	3.38	2.45	2.01	0.35
Stories about new technology	3.78	3.11	2.46	0.31
Factual science articles	3.28	2.77	2.57	0.18
Stories in which there is a rational explanation for everything	3.36	3.16	2.83	0.14
New-wave SF items				
Avant-garde fiction which experiments with new styles	3.38	3.29	3.46	0.00
Fiction based on the social sciences	3.92	3.65	3.68	0.09
Fiction that is critical of our society	4.13	4.12	4.72	−0.12
Fiction which deeply probes personal relationships and feelings	4.22	4.67	4.81	−0.21
Feminist literature	2.01	2.66	3.27	−0.26
Utopian political novels and essays	3.17	3.14	3.29	−0.03
Fantasy items				
Fantasy stories involving swords and sorcery	3.36	2.94	1.96	0.26
Stories about magic	3.58	2.99	2.91	0.18
Tales of the supernatural	3.40	3.34	3.23	0.03
Myths and legends	4.28	4.22	4.17	0.03

comparable GSS question, but this may reflect the fact that respondents had volunteered to answer a questionnaire focused on the space program.

The first row shows how strongly a general interest in science fiction correlates with support for the real space exploration program. I should note that when the dataset was published in software for student use, 1 was added to each of the preference ratings, to put them in a 1–7 scale rather than 0–6, to make it easier to handle on a computer keyboard. Here, the original 0–6 scale is used to make the numbers comparable to those in the study of science fiction convention goers. Scanning down the column for gamma correlations, we see a clear if complex picture. Liking science fiction in general, and liking stories about physical sciences and technology, are positively associated with support for the space program. Factual science articles and rationalist stories also show positive correlations, but much weaker ones. "Science" includes many fields unrelated to space exploration, and spaceflight in the context of the Challenger disaster may not have seemed very rational to many respondents.

Most of the New-Wave items show no relationships to support for the space program. Valuing personal relationships and Feminist literature reflect gender differences, so it is not surprising they correlate negatively with support for space funding. Or, they may suggest the values that contradict investment in spaceflight, which seems designed to escape the Earth rather than improve life here on Earth. The swords and sorcery subgenre, which is practically defined by *Star Wars*, was favored by SF fans more than traditional supernatural versions of fantasy, and here we see it correlating positively with spaceflight among a somewhat more representative sample of respondents. Clearly, the aspects of the science fiction subculture that are logically connected to spaceflight, really do correlate with support for it.

To replicate this general finding and seek departures from it, the other spaceflight items in S1986A were tabulated against preferences for the general science fiction item, results summarized below. For example, the first item below asked about putting civilians into space, such as Christa McAuliffe the schoolteacher who died in the Challenger disaster. Of the 306 respondents who felt this was important, the mean preference score for science fiction was 3.64. But among those who felt this was too dangerous, the mean was 3.05, rather lower. Among the 72 respondents who did not have a firm opinion, the mean for science fiction was 3.18, on the low side of the range but between the means for the two other responses.

Do you think that putting civilians into space is important-or is it too dangerous?

3.64 Important (306)
3.05 Too dangerous (79)
3.18 Don't know (72)

Some people say the United States should concentrate on unmanned missions like the Voyager probe. Others say it is important to maintain a manned space program as well. Which comes closer to your view?

2.58 Unmanned program only (31)
3.61 Manned as well as unmanned program (438)
2.53 No opinion (34)

Do you think the United States should build a permanently manned space station in orbit around the earth over the next few years or not?
3.97 Yes (279)
2.85 No (129)
2.83 No opinion (87)

Recently, there has been much talk about building a system of space satellites to defend us against nuclear attack. Do you think research on this idea should continue, or should research stop?
3.62 Research should continue (230)
3.33 Research should stop (246)
3.77 No opinion (22)

There has been much discussion about attempting to land people on the planet Mars. How would you feel about such an attempt—would you favor or oppose the United States' setting aside money for such a project?
3.82 Favor (244)
3.01 Oppose (172)
3.48 No opinion (86)

If you were asked to go along on the first rocket trip to the planet Mars, would you want to go or not?
3.92 Yes (283)
2.83 No (194)
3.78 No opinion (27)

Do you think we should attempt to communicate with intelligent beings on other planets, perhaps using radio?
3.90 Yes, definitely (233)
3.23 Yes, perhaps (188)
2.72 No (57)
3.14 No opinion (22).

In some of the lines above the number of cases is low, so random factors make the estimates uncertain, but in general liking science fiction does connect positively with expansion of space development.

7.5 The Analytical Laboratory

A remarkable, future-oriented but past-created dataset that can provide guidance on the meaning of science fiction during its Golden Age is the Analytical Laboratory from the central publication of that era, *Astounding Science Fiction*, renamed *Analog* in 1960 (Bainbridge 1980). From March 1938 through October 1976, stories in every issue of this magazine were rated in the "An Lab" readers' poll, developed by the editor, John W. Campbell, Jr. An incredible amount of fascinating literary data lies buried in the 464 polls that were published, covering twenty-five hundred fiction items. Half of these were short stories, and a third were novelettes. The remainder consisted of the most influential pieces of fiction, 70 "short novels" published whole in single issues, and 133 serialized novels published in a total of 370 installments. Since each installment was rated separately by readers, we will count them separately here. Included in these large numbers are many of the most popular works of science fiction ever written.

Let us start with a specific example. I have chosen the Lab for a very special issue, November 1949. Filled with stories by the greatest authors, this famous issue is the hero of its own science fiction story: The November 1948 issue contained a letter from Richard A. Hoen rating the stories in the November 1949 issue. In 2003, Andrew May blogged that this was a remarkable science fiction prophecy, and analyzed the similarities between the issue Hoen imagined and the one actually published a year later. Of course, this was neither coincidence nor prophecy, and Campbell had arranged for almost all of the predicted stories to be written by their authors. Hoen had predicted that *Gulf* would be written by Anson MacDonald, and Table 7.3 shows that it was actually written by Robert A. Heinlein. However, MacDonald was one of Heinlein's pen names. The symbolism of this unusual issue is dual: (1) Campbell believed that science fiction should indeed predict the real future. (2) Campbell also believed that the future of literature would involve its convergence with rigorous science.

The November 1949 issue of *Astounding Science Fiction* contained six stories, two of which were installments of novels by very popular writers. The readers sent letters—through regular mail because Internet would not exist until their distant future—ranking the stories from 1 (best) to 6 (worst). Campbell calculated the points by simply taking the average of the ranks across the readers who did the rating. The "Adjusted" column, as explained below, is an estimated place on a scale from 1 to 00 that corrects for the fact that issues differ in the number of stories they contain.

In the place, points, and adjusted scores, as in the game of golf, a low number is a good rating, while a high number is bad. This seems simple enough. But there are at least four reasons why we cannot blithely add and divide the place and point scores in an overall analysis of the authors and their twenty-five hundred stories. The first problem is that the Analytical Laboratory frequently fails to report votes on the least popular stories. In addition to the five items given point scores by Campbell in Table 7.3, the November 1949 issue also contained, "Finished." I added it to the list, putting it in sixth place, but there is no way to know how many points it received.

Table 7.3 The Analytical Laboratory for the November 1949 issue of *Astounding*

Place	Title	Author	Points	Adjusted
1	Gulf (Part 1)	Robert A. Heinlein	1.38	143
2	And now you don't (Part II)	Isaac Asimov	2.33	286
3	What dead men tell	Theodore Sturgeon	3.00	429
4	Final command	A. E. van Vogt	4.09	571
5	Over the top	Lester del Rey	4.90	714
6	Finished	L. Sprague de Camp	–	857

The second problem was mentioned by Campbell: "Not every reader letter casts votes on all the stories; thus the total number of votes cast for a particular story may not equal the total number of ballot letters." (Campbell 1943a) Probably, people will tend to skip stories they dislike. This means that the point scores for the least popular stories will be lower (better) than they deserve to be.

The third problem is that Campbell used an odd convention for expressing tie votes. For example, A. E. van Vogt won first place in the December 1948 issue, while Poul Anderson and Eric Frank Russell tied for second. In the Lab, Campbell gave second place to both Anderson and Russell, and awarded third place to a story by H. B. Fyfe. More properly, Fyfe should be in fourth place, since three stories got better ratings than his. Since Anderson and Russell were battling for second and third place, we should put each of them in "2.5" place. If many readers expressed tie scores the way Campbell did, then again some lower-rated stories would wind up with incorrectly good scores.

The fourth and most important problem comes from the fact that different issues contained different numbers of stories. Campbell recognized that this fact made it very difficult to compare from one issue to another. One time he commented, "The June issue carried seven stories besides the article; this means that point-score votes ranged from one to seven—and made point scores tend to run high. That's somewhat unfair, in a way—a third-place story or fourth-place story in such an issue has met and surpassed more competition, yet gets a tougher point score than the rearguard item in a five-story issue. Some day all things will be perfect—and a completely fair system of reporting may be worked out." (Campbell 1943b) The analysis here will use a specially-designed correction formula to defeat these four problems and make it possible to translate all scores to a single, uniform scale.

The place orderings, which exist for all 464 issues, can be converted to a uniform scale with a simple and mathematically sound formula (Bainbridge 1980). This was derived for me from probability logic by Toshio Yamagishi, an expert in preference research who was kind enough to contribute his expertise, and who has recently been publishing cross-cultural preference studies (Yamagishi et al. 2008). Suppose all twenty-five hundred stories were ranked from best to worst, in a single huge Lab. Now let Chance play the role of editor, selecting stories at random to fill the 464 issues. Finally, assume that stories within each issue were rated by a regular Lab, so we know which one is the most popular, which is second in the issue, and so on. Yamagishi derived a statistical formula that lets us predict the probability that a story in a given place in an issue of given size will come from any given level in the ranking of 2,500, here expressed as synthetic ranks from a hypothetical sample of 1,000 stories.

Notice that the An Lab has several possible units of analysis, including issues of the magazine, authors, stories, and story categories such as novel installments versus short stories. Over the years, Campbell mentioned several factors that might influence the popularity of a story, and once suggested that the second episode of a serial might have suffered because readers forgot characters and plot details over the month since the first episode (Campbell 1955). This highlights the possibility that later installments are less popular in general than the first installments, and my original research with this dataset did indeed show that first installments were rated higher than later ones.

Campbell commented on the length factor several times. "One of the problems inherent in science fiction is that each story actually is a brief glimpse of an alien world-scene. The longer the story, the more chance the author has to give a feel of reality—a texture of living fabric—to his world-picture. Result: a longer story, all things—and authors!—being equal, will have more satisfying effect for the reader." (Campbell 1956) Of course, it may simply be that readers best recall those stories that took longest to read, subconsciously multiplying the enjoyment experienced per page times the number of pages to arrive at a total impression. When an author writes a long novel he probably invests more effort in planning and characterization, so that even the first part of a long novel conveys more vivid images than an equally long segment of a shorter work.

Another time, Campbell explained: "Generally, the longer a story is, the more chance the author has to work out his background ideas, characters, and plotting. Serials generally take first place, primarily because the author can do a better job. Unlike here-and-now-stories, science fiction must describe even the common things of life—life in the story environment. More space gives more chance for that. The result is that there are very few long-remembered, 'classic' short stories, a few novelettes, but many much-mentioned serials." (Campbell 1946) To test this idea on all kinds of fiction, I tabulated place distributions for all 935 pieces of fiction published in the 187 five-story issues that contained no Lab ties, results show here in Table 7.4. The pattern is quite regular. Serials beat out short novels which surpass novelettes which win over short stories. Indeed, the short stories are crammed into the last three places.

Clearly, one level of aggregation at which preferences may be focused is that of the author. Table 7.5 gives An Lab data for the dozen highest-rated authors. A total of 53 authors published at least 10 items rated in the An Lab, and their ranks are based on the mean adjusted ranking place of each author's publications within its issue of the magazine. On February 9, 2014, I searched by each author's full name on Google, and report the number of hits listed in the table, except for two authors where at least one other prominent person had the same name. The final column of the table gives the Pearson's r correlations with hard-science SF from

Table 7.4 Kinds of fiction achieving each "place" in 187 *Astounding* 5-story issues

Place	Installments	Novels	Novelettes	Stories
1	70 %	51 %	20 %	2 %
2	18 %	40 %	42 %	5 %
3	10 %	9 %	24 %	22 %
4	2 %	0 %	11 %	33 %
5	0 %	0 %	3 %	38 %
Total	100 %	100 %	100 %	100 %
Number of items	145	35	294	461

Table 7.5 The top dozen authors from the Analytical Laboratory

Author	Items	Mean year	Mean lab rank		Google hits	Correlation with hard science SF
			Place	Points		
Robert A. Heinlein	35	1945	223	132	802,000	0.27
E. E. "Doc" Smith	13	1944	244	190	115,000	0.34
Jerry Pournelle	11	1973	280	265	437,000	0.29
A. E. van Vogt	59	1944	348	298	357,000	0.26
Harry Harrison	32	1966	321	316	Uncertain	0.23
Lawrence O'Donnell	11	1947	330	323	Uncertain	
Frank Herbert	28	1963	381	329	754,000	
Poul Anderson	67	1960	348	332	830,000	0.27
Hal Clement	29	1953	315	340	122,000	0.38
Jack Williamson	19	1944	348	343	330,000	0.38
Clifford D. Simak	39	1949	356	350	419,000	0.13
Isaac Asimov	45	1950	391	351	2,160,000	0.28

the 1978 study, for the 409 knowledgeable respondents, and for those of this dozen who were among the 27 authors who defined Factor I for the 276 best respondents. O'Donnell was not included in the 1978 study, and Frank Herbert tended to avoid definition in my correlational studies, given that his complex and multidimensional *Dune* series was originally published in *Analog*.

There should be no doubt that the Golden Age and subsequent decades of Astounding/Analog emphasized the hard-science dimensions of science fiction, and that its leading authors continue to be read and potentially to be influential.

7.6 Twenty Award-Winning Novels

Once the science fiction subculture had consolidated around the magazines, and the World Science Fiction convention was established as the main annual meeting, it was natural for there to be a system of awards comparable to the Oscars for movies. Called the "Hugo" after Hugo Gernsback, editor of the first SF magazine, this award came to have many categories, of which best novel of the year provides a good indicator of the subculture's orientation toward spaceflight. Table 7.6 lists the first twenty novels to receive the award, which was not given in 1954 or 1957, and was given to two novels with tie votes in 1966. Two columns at the right give the number of Amazon.com reader reviews as of February 2, 2014, and their mean score on a scale from 1 to 5. The Amazon.com reviews are not ideal as measure of popularity, because some reviewers evaluate the physical quality of the particular edition they bought, and some combine other books in a series such as *Dune*. Yet the number of reviews and mean ratings for these 20 Hugo winners do suggest the very different impacts these books continue to have today.

Table 7.6 The first 20 Hugo award-winning novels

Year	Author	Title	Reviews	Mean
1953	Alfred Bester	*The Demolished Man*	487	4.2
1955	Mark Clifton and Frank Riley	*They'd Rather Be Right (The Forever Machine)*	15	3.4
1956	Robert A. Heinlein	*Double Star*	86	4.2
1958	Fritz Leiber	*The Big Time*	38	3.4
1959	James Blish	*A Case of Conscience*	46	3.5
1960	Robert A. Heinlein	*Starship Troopers*	1,132	4.4
1961	Walter M. Miller, Jr.	*A Canticle for Leibowitz*	371	4.4
1962	Robert A. Heinlein	*Stranger in a Strange Land*	857	4.0
1963	Philip K. Dick	*The Man in the High Castle*	318	4.0
1964	Clifford D. Simak	*Here Gather the Stars (Way Station)*	777	4.4
1965	Fritz Leiber	*The Wanderer*	23	3.1
1966	Frank Herbert	*Dune*	1,745	4.4
1966	Roger Zelazny	*...And Call Me Conrad (This Immortal)*	41	4.0
1967	Robert A. Heinlein	*The Moon Is a Harsh Mistress*	379	4.5
1968	Roger Zelazny	*Lord of Light*	190	4.7
1969	John Brunner	*Stand on Zanzibar*	67	4.2
1970	Ursula K. Le Guin	*The Left Hand of Darkness*	269	4.2
1971	Larry Niven	*Ringworld*	248	3.8
1972	Philip José Farmer	*To Your Scattered Bodies Go*	99	4.1
1973	Isaac Asimov	*The Gods Themselves*	168	4.2

Earlier in this chapter we considered *The Demolished Man* by Alfred Bester, and *Mission of Gravity* by Hal Clement that to my mind was its chief competitor. In 2004, there was a special Retro Hugo competition to identify the best novel of 1954, and *Mission of Gravity* was a finalist. However, the award went to *Fahrenheit 451* by Ray Bradbury (Bradbury 1953). The novels by Bester and Bradbury represent high quality within the traditions of standard Literature, and explore some of the alternative ideologies common among people highly educated in the Humanities during the 1950s, not the hard sciences, and certainly not spaceflight. *Fahrenheit 451* has some of the qualities of George Orwell's *Nineteen Eighty-Four*, and may have been a reaction to the anti-intellectualism of the so-called McCarthy Era in American Politics (Orwell 1949; Hofstadter 1963). The title refers to the supposed temperature at which paper burns, and book-burning is the central technology of the novel.

The Clifton and Riley novel that won the Hugo in 1955 is somewhat controversial within the science fiction community, and is not regarded as a great work. The chief theme has nothing to do with spaceflight, but postulates a technology that can restore youth and beauty at the price of removing a person's intellectual individuality. Originally published in *Astounding Science Fiction*, it is sometimes linked to the birth of Scientology, a real-world religious movement that offers technologies for mental improvement, and Scientology was originally launched as Dianetics from the pages of Astounding in 1950 (Lewis 2009). One can also see parallels with the pressures for intellectual conformity of McCarthyism and other trends in American culture of the 1950s, that were overturned in the rebellious 1960s.

In 1956, Robert A. Heinlein won the first of his four Hugos for novels, all four of which concern spaceflight. However, each uses space as the canvass on which to paint human stories with political or religious implications, and each, to mix metaphors slightly, is an ideological tapestry. Given that Heinlein was the most highly regarded author of his period to promote real spaceflight vigorously, the fact the he uses space to frame stories with frankly radical political implications is remarkable. He has often been decried as a right-wing radical, yet this charge is unfair (Panshin 1968). Rather, there were three reasons for Heinlein's apparent ideological position. First, he sought to compose interesting stories, which in the intellectual context of mid-twentieth century America required exploring unusual perspectives. Second, like many others in the science fiction community he really was a political radical, but his viewpoint does not map easily onto the traditional left–right spectrum. Third, as an intellectual he wanted to explore hypothetical alternatives to every-day thinking, but in the realization that his fiction could not really test what impact the ideas might have if they were applied in the real world.

The very beginning of *Double Star* seems highly critical of spaceflight, suggesting that astronauts are arrogant and perhaps a bit stupid, lacking the proper sense of human interaction that a competent member of society is supposed to have:

> If a man walks in dressed like a hick and acting as if he owned the place, he's a spaceman. It is a logical necessity. His profession makes him feel like boss of all creation; when he sets foot dirtside he is slumming among the peasants. As for his sartorial inelegance, a man who is in uniform nine tenths of the time and is more used to deep space than to civilization can hardly be expected to know how to dress properly. He is a sucker for the alleged tailors who swarm around every spaceport peddling "ground outfits." (Heinlein 1956)

However, these words do not belong to Heinlein, but to his protagonist, and the novel is written as a first-person narrative. This scornful view of spacemen belongs to The Great Lorenzo, a down and out actor who stumbles into the bizarre gig of impersonating a politician who has been kidnapped and injured, in an upcoming election. Lorenzo not only pretends to be a different person, but to hold the views of that person, which are very different from his own. In particular, the politician was in favor of equal political rights for Martians, which Lorenzo personally opposes. Indeed, one can interpret the novel as a parody of people who do not favor the values of spaceflight, implying that they are selfishly concerned with manipulating the people around them to their own benefit, and lack any wider moral values.

Heinlein's *Starship Troopers* is especially controversial, because it seems to promote non-democratic political theories, and a major motion picture exploits it as a high-art parody of Fascism (Heinlein 1959). The novel depicts the heroic military adventures of young people in an interstellar war against Bugs who are formidable insect enemies despite apparently lacking individual consciousness and artificial technologies. The most instructive way to conceptualize *Troopers* is not as propaganda but speculation, and indeed Heinlein preferred to call his genre not *science fiction* but *speculative fiction*. How can individuals develop loyalty to each other? How is it possible to maximize both individual and collective responsibility? One feature of the future society described in *Troopers* is that only people who have served honorably in the military have the political right to vote. Whatever one thinks of this as a political principle, it raises the issue of whether rights should be earned rather than God-given. By staging the drama in outer space, Heinlein not only postulates an implacable alien enemy, but also develops the theme of human maturation, as young people who voyage far from Earth are symbolically leaving home to seek adult destinies.

Stranger in a Strange Land also explores the theme of maturation in a speculative manner, because it can be considered to be a sequel to one of Heinlein's juvenile novels, *Red Planet* (Heinlein 1961, 1949). That 1949 book for boys concerns liberation from a oppressive boarding school, exploration of an alien environment, and discovery that a Martian creature used as a pet was actually the juvenile stage of a highly intelligent exotic species of humanoid. *Stranger* does not exactly fit the assumptions of *Red Planet*, but concerns a man who returns from Mars after having been instructed by these Martians and begins to share their alien philosophy with Earthlings. The motto of the book is *grok*, a form of harmony between people that is deeper than mere cognitive understanding, and may have been a trope on "grow close." This theme harmonized with the counterculture of the 1960s and expresses the fundamental ideal of communal movements, some of which became popular in the real world during the 1960. As it happens, I have written books on the basis of a close study of two radical religious communes that grew out of the counterculture of the 1960s, The Process and The Family, both of which practiced group marriage and possessed ideologies that could have derived from alien cultures (Bainbridge 1978, 2002).

In many respects, *The Moon is a Harsh Mistress* is the ideological opposite of *Stranger in a Strange Land*, extolling individualist rather than communal values (Heinlein 1966). It explores the meaning of Earth's satellite as a metaphor for personal autonomy, because its separation from our planet represents the distance an individualist seeks from the mass of humanity. Its motto, TANSTAAFL, has some affinity with the idea in *Starship Troopers* that political rights must be earned: "There Ain't No Such Thing As A Free Lunch." This theme in Heinlein's work links him to modern Libertarians, to the non-traditional components of the Tea Party within the Republican Party, and to a few other authors, notably Ayn Rand who sanctified individualism in *The Fountainhead* and *Atlas Shrugged* (Rand 1943, 1957).

Several other novels in the list of twenty early Hugo winners also use the astronomical realm as if it were a multi-dimensional form of graph paper, on which to map alternative ideologies. *A Case of Conscience* by James Blish explores the contradictions when a Jesuit priest contemplates an alien culture that has admirable morality without the need of any religion (Blish 1958). One could argue that Frank Herbert's hugely impactful novel *Dune* contemplates environmentalism and the decline of empires under inspiration from Islamic culture (Herbert 1965).

Roger Zelazny's *Lord of Light* explicitly draws upon Hinduism (Zelazny 1967). *The Left Hand of Darkness* by Ursula K. Le Guin considers the anthropological implications of gender, on a planet where sexual identity is not innate but arises only temporarily when a child is being created (Le Guin 1969). Larry Niven's *Ringworld* was the beginning of a series having many aspects, but fundamentally returns to the hard-science tradition by postulating an artificial planet in the form of a rotating ring around its star, millions of miles across (Niven 1977).

The Gods Themselves by Isaac Asimov was somewhat of a departure from his standard methods, highlighting an alternate reality rather than our own galaxy, and involving aliens with exotic biology and psychology (Asimov 1972). His most influential writing, within the science fiction genre, was completed much earlier than 1973 when he won the Hugo for best novel, and even before the Hugos existed. In 1966, the Hugo for "Best All-Time Series" was awarded to Asimov's original *Foundation* trilogy, which concerned the fall of a future galactic civilization founded by humans who never encountered aliens and which was originally published in *Astounding*. Of the four competing finalists for that special Hugo, three emphasized spaceflight: Heinlein's "future history" series, the novels about Mars by Edgar Rice Burroughs, and the "space opera" action and high-tech adventure series about an elite spaceman corps of Lensmen by E. E. "Doc" Smith. The other competitor in 1966 was the high fantasy *Lord of the Rings* tales by J. R. R. Tolkien.

7.7 A Variable Time Machine

Many early science fiction stories, probably most up until the dawn of the Golden Age, were set in the present or the very near future. But once the genre became well established, most stories moved into the far future, to render plausible very major social and technological differences from the world the readers inhabited. Hugo Gernsback, most influential at the end of the 1920s, felt that science fiction should serve revolutionary but near-term goals: to educate readers in scientific principles, to inspire young people to enter technical professions, to sketch future inventions which readers could then perfect, and to generate enthusiasm for science and technology among the general public. There is evidence that all these effects did follow in the early days; for example, the great social movement which produced modern space rocketry was inspired by science fiction classics like those discussed above.

The specific branch of technology most vigorously promoted by Gernsback was not spaceflight but television, for example in his own novel *Ralph 124C41+*, and in the pages of popular electronics magazines which he edited (Gernsback 1925). Ever since then, electronics has remained the area of technological innovation most open for accomplishments by individuals and small teams, most recently in computer software and Internet-based social media. Yes, science fiction has contributed some inspiration, for example through novels like *Neuromancer* by William Gibson in 1984 and *Snow Crash* by Neal Stephenson in 1992 that publicized ideas like cyberspace and avatars (Gibson 1984; Stephenson 1992). Yet those technologies were developing rapidly anyway, so it is hard to assess how significant such cyber-SF was. Most importantly, no sane person is trying to build a spaceship in their back yard, yet much new information technology does arise from very modest but imaginative projects.

Today there is real question whether science fiction literature still promotes scientific and technological values, or whether it now speaks in many contradictory voices, some adamantly opposed to the ideology of its founders. In the 1960s, the New-Wave movement in science fiction promoted avant-garde literary experimentation, criticism of technology, and an interest in the social rather than in the physical sciences. A large and popular group of Fantasy authors found a secure place with the science fiction field, mainly producing what is called Sword-and-Sorcery. Ignoring science, these writers spin stories around magic and swordplay, implicitly urging technological primitivism and a reversion to our barbarian ancestors. Furthermore, whatever the ideological thrust of SF literature, there is some question whether it gets communicated successfully to the mass public, because popular visual media may present only the most distorted and attenuated shadow of science fiction.

References

Asimov, I. (1972). *The gods themselves*. Garden City, New York: Doubleday.

Bainbridge, W. S. (1976a). *The spaceflight revolution* (p. 201). New York: Wiley-Interscience.

Bainbridge, W. S. (1976b). *The spaceflight revolution* (p. 206). New York: Wiley-Interscience.

Bainbridge, W. S. (1976c). *The spaceflight revolution* (p. 200). New York: Wiley-Interscience.

Bainbridge, W. S. (1978). *Satan's power: A deviant psychotherapy cult*. Berkeley: University of California Press.

Bainbridge, W. S. (1980). The analytical laboratory, 1938–1976. *Analog, 100*(1), 121–134.

Bainbridge, W. S. (1986). *Dimensions of science fiction*. Cambridge, Massachusetts: Harvard University Press.

Bainbridge, W. S. (2002). *The endtime family: Children of God*. Albany, New York: State University of New York Press.

Bainbridge, W. S. (2012). The Psychoanalytic movement. In W. S. Bainbridge (Ed.), *Leadership in science and technology* (pp. 520–528). Thousand Oaks, California: Sage.

Bester, A. (1953). *The demolished man*. Chicago: Shasta.

Blish, J. (1958). *A case of conscience*. New York: Ballantine.

Bradbury, R. (1953). *Fahrenheit 451*. New York: Ballantine.

Burroughs, E. R. (1917). *A princess of Mars*. Chicago: A. C. McClurg.

Campbell, J. W. (1943a). The analytical laboratory. *Astounding Science Fiction*, October, 29.

Campbell, J. W. (1943b). The analytical laboratory. *Astounding Science Fiction*, September, 48.

Campbell, J. W. (1946). The analytical laboratory. *Astounding Science Fiction*, July, 122.

Campbell, J. W. (1955). The analytical laboratory. *Astounding Science Fiction*, June, p. 118.

Campbell, J. W. (1956). The analytical laboratory. *Astounding Science Fiction*, June, 72.

Clement, H. (1954). *Mission of gravity*. Garden City, New York: Doubleday.

de Bergerac, C. (1899). *A voyage to the Moon*. New York: Doubleday and McClure.

Dozois, G. (2013). Summation: 2012. In G. Dozois (Ed.), *The year's best science fiction* (p. xvii). New York: St. Martin's Griffin.

Gernsback, H. (1925). *Ralph 124C41+, a romance of the year 2660*. Boston, Massachusetts: Stratford.
Gibson, W. (1984). *Neuromancer*. New York: Ace.
Heinlein, R. A. (1949). *Red planet, a colonial boy on Mars*. New York: Scribner's.
Heinlein, R. A. (1956). *Double star* (p. 1). Garden City, New York: Doubleday.
Heinlein, R. A. (1959). *Starship troopers*. New York: Putnam.
Heinlein, R. A. (1961). *Stranger in a strange land*. New York: Putnam.
Heinlein, R. A. (1964). On the writing of speculative fiction. In L. A. Eshbach (Ed.), *Of worlds beyond* (p. 13). Chicago: Advent.
Heinlein, R. A. (1966). *The moon is a harsh mistress*. New York: Putnam.
Heinlein, R. A. (1969). Science fiction: Its nature, faults, and virtues. In B. Davenport (Ed.), *The science fiction Novel* (p. 15). Chicago: Advent.
Herbert, F. (1965). *Dune*. Philadelphia: Chilton.
Hirsch, W. (1958). The image of the scientist in science fiction. *American Journal of Sociology, 63*(5), 506–512.
Hofstadter, R. (1963). *Anti-intellectualism in American life*. New York: Knopf.
Lait, J., & Mortimer, L. (1953). Mars confidential. *Amazing Stories, 27*(April–May), 19.
Lasswitz, K. (1896). Nature and the individual mind. *The Monist, 6*(3), 401–402.
Lasswitz, K. (1969). *Auf zwei Planeten*. Frankfurt: Scheffler.
Le Guin, U. K. (1969). *The left hand of darkness*. New York: Walker.
Lewis, J. R. (Ed.). (2009). *Scientology*. New York: Oxford University Press.
Milburn, C. (2010). Modifiable futures: Science fiction at the bench. *Isis, 101*(3), 560–569.
Niven, L. (1977). *Ringworld*. New York: Holt, Rinehart and Winston.
Orwell, G. (1949). *Nineteen eighty-four*. New York: Harcourt, Brace.
Panshin, A. (1968). *Heinlein in dimension, a critical analysis*. Chicago: Advent.
Patch, H. R. (1922). The tradition of the goddess Fortuna in Roman literature and in the transitional period. *Smith College Studies in Modern Languages, 3*(3), 166.
Rand, A. (1943). *The fountainhead*. Philadelphia: Blakiston.
Rand, A. (1957). *Atlas shrugged*. New York: Random House.
Serviss, G. P. (1911). *A Columbus of space*. New York: Appleton.
Serviss, G. P. (2009). *Edison's conquest of Mars* (pp. 6–7). Mumbai, India: Tutis Digital Publishing.
Smith, E. E. (1946). *The skylark of space: The tale of the first inter-stellar cruise*. Cranston, Rhode Island: Southgate.
Stephenson, N. (1992). *Snow crash*. New York: Bantam.
Verne, J. (1869). *From the earth to the moon*. Newark, New Jersey: Newark.
Wells, H. G. (1898). *The war of the worlds*. New York: Harper.
Wells, H. G. (1901). *The first men in the moon*. London: G. Newnes.
Yamagishi, T., Hashimoto, H., & Schug, J. (2008). Preference versus strategies as explanations for culture-specific behavior. *Psychological Science, 19*(6), 579–584.
Zelazny, R. (1967). *Lord of light*. Garden City, New York: Doubleday.

Chapter 8
Media

Television and motion pictures that depict spaceflight have great power to shape public conceptions of travel between the planets and stars, despite audience awareness that the technologies shown are the spurious result of special effects technologies, rather than being representations of reality. A very few films have attempted to be realistic in all senses of the term, but they are greatly outnumbered by others that are fantastic in violating the laws of science while projecting exciting images. In addition, the relative popularities of different kinds of spaceflight films speak volumes about what travel beyond the Earth already means to American audiences.

This chapter will organize much of the discussion around the equivalent of questionnaire survey data, the ratings given by tens of thousands of people to 80 spaceflight movies in the Netflix recommender system, augmented by data from the Internet Movie DataBase (IMDB). These quantitative data will make the most sense only after an historical overview sets the stage. I have viewed most of these 80 films, and all of the others discussed here, so observation was a research methodology at least equal in importance to statistical analysis in developing this chapter.

It is worth pointing out that movies and television are themselves advanced technologies, from the perspective of the full sweep of cultural evolution, and that these two media have interacted in complex ways with each other. Today they seem to be merging as theaters convert to digital projection systems, and movies are delivered on demand to any home in advanced societies.

8.1 Spaceflight in Movies and Television

The very first spaceflight movie, *Voyage dans la Lune (A Trip to the Moon)* dates from 1902 and was arguably the most stunning early demonstration of special effects in cinema. As Wikipedia rightly explains, its creator, Georges Méliès, "was a French illusionist and filmmaker famous for leading many technical and

W.S. Bainbridge, *The Meaning and Value of Spaceflight*,
Space and Society, DOI 10.1007/978-3-319-07878-6_8

narrative developments in the earliest days of cinema. Méliès, a prolific innovator in the use of special effects was one of the first filmmakers to use multiple exposures, time-lapse photography, dissolves, and hand-painted color in his work. Because of his ability to seemingly manipulate and transform reality through cinematography, Méliès is sometimes referred to as the first 'Cinemagician'". This roughly 15-min film was inspired by the two most prominent prior lunar voyage novels, *De la Terre à la Lune* by Jules Verne and *The First Men in the Moon* by H. G. Wells, using Verne's huge gun to get there and basing the alien inhabitants on Wells (Verne 1869; Wells 1901).

The most significant spaceflight movie, the one that did the most to advance the real technology, was *Frau im Mond* (*Woman in the Moon*). Although released in 1929, two years after *The Jazz Singer* really launched "talking pictures" in America, *Frau im Mond* was a silent film. It was produced in Germany, which had suffered terrible economic crises in the 1920s, and even if it had been feasible to make it as a sound film, not many of the German movie theaters would have been able to present it in that manner. Unfortunately, the anachronism that *Frau in Mond* was advanced in terms of spaceflight technology, but retarded in terms of motion picture technology, limited its impact worldwide. It was produced and directed by Fritz Lang, the highly acclaimed cinema creator, whose earlier work, *Metropolis*, was also a science fiction epic, artistically analyzing the negative meanings of robots and other technological innovations for workers, and suggesting a reformulation of capitalism that might render it benign for humanity. *Frau im Mond* was superficially based on a novel by Lang's wife, Thea von Harbou, but more profoundly on the space technology ideas of rocket pioneer Hermann Oberth.

The plot of *Frau in Mond* is not memorable, merely an excuse to send people to the Moon. When they get there, the landscape looks lunar but there is an atmosphere that allows the characters to walk around in ordinary clothing. However, the means for getting there is technically both advanced and accurate. Already in his 1923 treatise, *Die Rakete zun den Planetenräumen*, Oberth had explained in some detail how liquid-fuel rockets could be built, and using the multi-stage principle travel to other worlds. He was intimately involved in the movie production. The film provided a propaganda boost for the Verein für Raumschiffahrt, the German spaceflight society that was developing real liquid-fuel rockets and would contribute personnel for the Peenemünde project that developed the V-2 (Ley 1951).

The earliest noteworthy American movie to depict spaceflight was visionary but not technically accurate. Titled *Just Imagine*, this musical comedy was produced in 1930 and imagined how the world might be in 1980. I first saw the film in that year, during a 50th anniversary event at the University of Washington, and it boggled my mind to compare fantasy with reality. New York, according to *Just Imagine*, would be a city of even greater skyscrapers than it really became, with people flying their private airplanes between them, rather than merely driving cars. The flight to Mars is achieved on a small rocket plane, that carries only a tiny fraction of the fuel that actually would be required, and Mars proves to be a farce inhabited by crazy entertainers.

There could hardly be greater contrast with the 1950 American spaceflight movie, *Destination Moon*, which like *Frau im Mond* presented serious technical ideas and will be considered in some depth in a later section of this chapter. Similarly, at the same point in spaceflight history, that new medium, television, was also used as an effective medium for propaganda and even education about spaceflight. At the dawn of broadcast TV, three popular adventure programs prepared children for the coming space age: *Captain Video* (1949–1955), *Space Patrol* (1950–1955), and *Tom Corbett*, Space Cadet (1950–1955). I watched most early episodes myself when they were broadcast, then viewed a number around 1990 when they were available on mail-order videocassettes that were probably pirated. Unfortunately, very few episodes of *Captain Video* are currently available, because it was broadcast on the DuMont network which went out of business in 1956, so apparently copies no longer exist (Weinstein 2004).

The programs depicted spaceflight in a manner that was certainly not entirely realistic, but did communicate some concepts that would serve as stepping stones in the viewer's education toward solid knowledge of astronomy and physics. All three took place primarily within the solar system, so they avoided dubious concepts like warp drive or hyperspace jumps, but they depicted journey times between planets as being only a few hours, in order to permit rapid progress through the scenes of the story.

For example, the episode of *Space Patrol* broadcast November 7, 1950, depicts the launch of a Patrol spaceship piloted by Commander Buzz Corey, in the company of his young lieutenant, Cadet Happy. They are at Space Patrol headquarters on the man-made planet, Terra, in the thirtieth century, and have just learned that their nemesis, the evil Count Baccarratti has escaped in a stolen Patrol ship. The typewritten script for that episode depicts Buzz and Happy taking off after him in their own ship:

FADE IN SLIDE

ANNCR: (CUE) And now back to the adventures of SPACE PATROL!
SOUND: (CUE) ROCKET WHOOSH... JETS AND DYNAMO HUM BG [background]
DISSOLVE: TO COCKPIT
BUZZ: (PILOT) All set, to blast off, Hap?
HAPPY: Yes sir.
BUZZ: Here we go. (ADVANCES THROTTLES) And we're on our way to Saturn.

CAMERA PEDESTAL DOWN

HAPPY: Say, Commander. Just how much of a start does Baccarratti have?
BUZZ: Quite a bit, Happy. He will already have landed *somewhere* by the time we get there... even though this is a faster ship.
HAPPY: How'll we ever locate him, sir? All we know is that he left here on the Saturn orbit. And that doesn't necessarily mean that he'll go to Saturn.
BUZZ: That's right, Hap. But I have alerted the Saturn Space Patrol Unit number 83. They'll be starting out from Saturn on the Terra orbit.

HAPPY: Good, sir. Maybe they can pick him up… or at least get his position located before we get there.

BUZZ: That's the idea, Hap. And I hope they don't miss him. If he should head for one of Saturn's nine moons… we might never find him.

HAPPY: Gee, sir. We might be trampin' around the mountains and rocks of some barren satellite for weeks… looking for that guy (Script provided by Nina Bara, the actress who played Baccarratti's accomplice in this episode and Norman Jolley is listed as the writer, a kinescope date is given as November 3 1950).

It is worth adding that ships in *Space Patrol* took off horizontally and had big front windows like a bus windshield, actually lacking glass so we could see the crew clearly and without reflections from studio lights, as they sat in the same manner as the pilot and co-pilot of a traditional airliner. The reference to "camera pedestal down" meant that by lowering the television camera, the ship would appear to rise, as if it were taking off.

The discussion of orbits is partially realistic, because it gets the children thinking about the curved path through space needed to voyage to another planet, and it explains that Saturn has moons that might be alternate destinations. Nine was indeed the number of moons known to astronomers at the time. The program does not however depict flights in which the engines shut off after a few minutes, which would be most consistent with the idea that there was just one orbit to Saturn—an orbit technically of the Sun. We can then wonder how different the paths of the two ships would be, given that they are capable of different speeds. Furthermore, the least-energy orbit *to* Saturn is not at all the same as the least-energy orbit *from* Saturn, and the program seems to conceptualize them like two lanes of a highway that follow the same path but in opposite directions. *Space Patrol* does not depict the crew going through high-G acceleration, but merely taking off as an aircraft would.

Ships in *Tom Corbett*, Space Cadet launched vertically, with pilot and co-pilot facing the camera through a window not unlike the one in *Space Patrol*, but placed on the side of the ship. Tom usually operated the ship, sitting next to his mentor, Captain Strong, while the two other crew members were on other decks of the ship, one above and one below the command deck. A launch did subject the crew to powerful forces of acceleration, and in rare episodes they were depicted floating around in zero gravity during the flight, either inside or outside the vessel. However the special effects were very difficult to do at that time in the history of television, so a dramatic convention such as magnetic boots allowed the crew to walk around normally.

The show was based on the 1948 juvenile novel *Space Cadet* by Robert A. Heinlein, and had as its technical advisor Willy Ley, who had been active in the Verein für Raumschiffahrt before emigrating from Germany to the United States where he became a writer of popular science books and articles (Heinlein 1948). Fittingly, the last episode broadcast June 25, 1955, was a final examination for the cadets, leaving open the question of whether a new semester of their education to become astronauts would follow, or the sequel would be the real space program.

Captain Strong gave them this assignment: "You will fly to Mars and return on an orbital course of your own choosing. You will be graded on elapsed time, distance covered, fuel consumed, condition of ship on arrival, adherence to all space traffic rules and general performance".

8.2 Sets of Related Films

While a successful television series like *Space Patrol* comprises hundreds of episodes, movies are brief and thus may have relatively little impact on the audience, unless they are connected into a series. A very successful series becomes a franchise encompassing other media. This was true for *Captain Video*, *Space Patrol*, and *Tom Corbett*. In childhood I owned a cardboard *Captain Video* spaceship control panel, a working *Space Patrol* monorail train, and a model of Tom Corbett's academy complete with tiny figurines. Today I have recordings of 102 *Space Patrol* radio dramas spanning the period 1952–1955, and a *Tom Corbett*, Space Cadet thermos bottle.

The first space-oriented franchise to appear in the cinema was *Flash Gordon*, and Table 8.1 lists data about its three movies, plus *Buck Rogers*. These were competing outer space franchises which emerged from different quadrants of science fiction literature but were primarily known for their comic strip manifestations in newspapers. I placed them together in the table, because the four movies were produced by the same team with the same hero, swimming champion Buster Crabbe playing the roles of both Flash Gordon and Buck Rogers.

Flash Gordon was derived without benefit of author's permission from the Mars novels by Edgar Rice Burroughs, and was the primary influence on the later *Star Wars* franchise. *Buck Rogers* evolved from a novel by Philip Francis Nowlan published in *Amazing Stories*, the first science fiction magazine, back in 1928 (Nowlan 2010). *Flash Gordon* was in the action-adventure genre of science fiction, nearly sword-and-sorcery and indeed including examples of both swords and magic that may or may not have been based on secret science. *Buck Rogers* also emphasized action, but had more affinity with intellectual forms of science fiction, including emphasis on the invention of fantastic new technologies. All four movies listed in the table involved spaceflight from Earth to another planet, and were structured as episodic serials, each in total rather longer than an ordinary movie.

Table 8.1 begins with the title and year of release of the movie, then reports data from the preference ratings from the Internet Movie DataBase and Netflix. Anyone who wishes may check the IMDB for descriptions of all 80 spaceflight movies covered in this chapter, including the most recent statistics on ratings by registered users of the archive. Users were asked to rate movies they had seen on a preference scale from 1 to 10, and the fact that only registered users could vote allowed the system to prevent counting multiple votes from any one person. The numbers reported here were collected prior to September 2012. At that time, data from 1,028 IMDB users gave *Flash Gordon* a 7.3 rating on the 1–10 scale. Checking again on December 15,

Table 8.1 Five spaceflight movie traditions

Title	Year	IMDB rating	IMDB raters	Netflix mean	Netflix raters	IMDB rank	Netflix rank
Flash Gordon	1936	7.3	1,028	3.10	613	14.5	53.0
Flash Gordon's Trip to Mars	1938	7.3	475	3.17	470	14.5	50.0
Buck Rogers	1939	7.2	432	2.96	539	19.5	63.0
Flash Gordon Conquers the Universe	1940	6.9	632	3.13	427	25.5	52.0
2001: A Space Odyssey	1968	8.4	222,378	3.80	57,445	5.5	11.5
2010	1984	6.7	25,952	3.50	9,661	30.5	31.0
Solaris	1972	8.0	28,784	3.24	5,890	7.0	46.0
Solaris	2002	6.2	45,779	2.50	26,444	53.5	78.0
Alien	1979	8.5	277,717	4.11	48,496	3.5	5.0
Aliens	1986	8.5	255,563	4.14	43,518	3.5	4.0
Alien 3	1992	6.4	116,367	3.53	14,336	46.0	30.0
Alien: Resurrection	1997	6.2	97,440	3.48	15,828	53.5	33.0
Pitch Black	2000	7.1	100,508	3.57	44,178	23.0	24.0
The Chronicles of Riddick	2004	6.5	94,464	3.56	51,263	40.0	25.0

2013, we find that 1,241 users give a rating of 7.2, so over time as more and more people rate a film the rating may change slightly. The 7.3 and 7.2 are not simple mean ratings, but weighted averages using procedures that IMDB refuses to reveal fully, because they were designed to prevent people from gaming the system to inflate ratings for a film in which they have a vested interest.

IMDB reports that the full running time of *Flash Gordon* is 245 min, and classifies it as action, adventure, and sci-fi. Users contribute not only ratings but written commentaries, and the IMDB page for each film features a synopsis by one of them. The synopsis of *Flash Gordon* appears to have been written by one of the world's leading arachnologists, and reads:

> A rogue planet is 'rushing madly toward the earth.' Impending doom creates worldwide pandemonium. But maverick scientist Dr. Zarkov hopes to stay disaster by travelling to the new planet in his experimental rocket. Two chance-met strangers, athletic Flash Gordon and damsel in distress Dale Arden, go with him. Arrived, the trio find Mongo to be a planet of wonders, warring factions, and deadly perils, its orbit controlled by Emperor Ming who has his own sinister plans for earth. Can our heroes, armed only with science and sex appeal, stop him.

The Netflix data were obtained in 2006, and were the dataset used in a prominent competition that year that challenged competitors to devise the best algorithm to predict ratings of a movie by a user based on ratings of other movies by that user calibrated against ratings of the movie and some of those other films by other raters. The goal of the Netflix rating system is to tailor advertising to users so they will rent movies from Netflix that they will like, thus encouraging them to follow the advice of Netflix in future rentals. Having the raw data, which I do not have in the case of the IMDB, allows sophisticated statistical analysis, which we will examine later in this chapter. Netflix uses a preference rating scale of 1–5, rather than 1–10 as in the case of IMDB, and here we see that 613 Netflix raters gave *Flash Gordon* a mean score of 3.10.

Because the two data sources used different scales, it is difficult to compare them directly, so the last two columns of the tables here give a movie's ranks in the full set of 80 spaceflight films. In the case of IMDB, the ratings were given only to one decimal point, so many films tied. When the mass media report rankings, they often make mistakes in their reporting, as the previous chapter noted about the polls in *Astounding Science Fiction*. For example, in reporting ranks for films, two of which were tied for second place, they might give the ratings as: 1, 2, 2, 3, 4. This is obviously wrong, because the film listed in 4th place is really in 5th. Better would be: 1, 2.5, 2.5, 4, 5, so that is the system used here.

After the *Flash Gordon* and *Buck Rogers* series, Table 8.1 lists the historically and artistically significant *2001: A Space Odyssey* and its less important but still respectable sequel, *2010*. Both films assert that spaceflight has transcendental meaning, symbolized by mysterious contact with advanced extraterrestrial intelligence, but both also strove to be technically accurate. Notably, *2001* recognized that interplanetary voyages would take a long time, and artificial gravity could be achieved only by rotating a wide, toroidal crew compartment (Agel 1970). The climax of *2010* combines advocacy for international cooperation with a plausible physics lesson by having a Russian spacecraft serve as the first stage, and an American spacecraft as the second stage, of a two-stage solution to the problem of how to accelerate the combined crews rapidly enough to escape disaster.

The 2002 *Solaris* was an American remake of the 1972 Russian version, and elsewhere I have explained that the fundamental meaning of these films may not really be about spaceflight at all (Bainbridge 2014a). Rather, the mental distortions imposed on human space explorers by an apparently intelligent planet may really have been a metaphor for the influence of the Soviet Union on Poland, in the original novel by symbolist Polish writer, Stanislaw Lem. The high-quality *Alien* series depicts the planets of other stars in sinister terms, and are horror films in which the monsters are extraterrestrials. Optionally, they can be viewed as artistic and even political critiques of high technology generally, and perhaps spaceflight specifically. The fact that the first two *Alien* films get higher ratings than *2001* in both datasets may reflect the obsolescence of some of the themes in *2001*, or simply the emotional impact of well-made horror films. Elsewhere I explored the *Riddick* mythos extensively, as it was manifested in two computer games as well as the two movies, and in 2013 a third film was released (Bainbridge 2011). A key theme is whether complete immorality may be more ethical than distorted morality, in a universe where spaceflight facilitates the unrestrained spread of evil.

I have also examined the *Star Trek* and *Star Wars* franchises in great depth elsewhere, and they are familiar to everyone, so they will receive only brief consideration here. In the IMDB data, as shown in Table 8.2, the first two *Star Wars* films are tied for first place, among the 80 spaceflight films, and hold the first two places

Table 8.2 *Star Wars* and *Star Trek* films

Title	Year	IMDB mean	IMDB raters	Netflix mean	Netflix raters	IMDB rank	Netflix rank
Star Wars: A New Hope	1977	8.8	457,917	4.50	85,184	1.5	2.0
Star Wars: The Empire Strikes Back	1980	8.8	407,841	4.54	92,471	1.5	1.0
Star Wars: Return of the Jedi	1983	8.4	318,208	4.46	88,846	5.5	3.0
Star Wars: The Phantom Menace	1999	6.5	280,640	3.60	71,143	40.0	21.0
Star Wars: Attack of the Clones	2002	6.7	234,581	3.55	92,671	30.5	26.5
Star Trek: The Motion Picture	1979	6.3	35,852	3.43	20,487	50.5	35.5
Star Trek II: The Wrath of Khan	1982	7.7	52,647	3.83	37,880	9.0	10.0
Star Trek III: The Search for Spock	1984	6.6	33,158	3.60	23,764	35.0	19.5
Star Trek IV: The Voyage Home	1986	7.2	35,926	3.71	23,342	19.5	14.0
Star Trek V: The Final Frontier	1989	5.2	26,498	3.32	24,190	65.5	45.0
Star Trek VI: The Undiscovered Country	1991	7.2	33,071	3.63	22,170	19.5	16.0
Star Trek: Generations	1994	6.5	36,623	3.61	20,663	40.0	17.5
Star Trek: First Contact	1996	7.5	62,599	3.77	35,367	11.0	13.0
Star Trek: Insurrection	1998	6.3	35,332	3.55	26,927	50.5	26.5
Star Trek: Nemesis	2002	6.4	36,873	3.44	36,519	46.0	34.0

in the Netflix data. The Netflix data included only five of the six films shown in theaters as two trilogies, and the missing one is generally considered better than the two other members of its trilogy. But there is no doubt the first trilogy was better than the second, more innovative and more coherent. The popular press has suggested that the planet-destroying Death Star that provides the climax for the first and third films was based on spaceflight pioneer Hermann Oberth's ideas for a space weapon that would focus the rays of the sun on enemy targets. But this reflects the ignorance of popular writers. The first *Star Wars* film is filled with clear references to the 1938 serial *Flash Gordon's Trip to Mars*, and it features a space gun Ming the Merciless was firing from Mars to destroy the Earth.

As *Star Wars* was inspired by *Flash Gordon*, which was in turn inspired by the Mars novels of Edgar Rice Burroughs, *Star Trek* was an adult version of *Captain Video*, *Space Patrol*, and *Tom Corbett*, Space Cadet. Like them, it was connected to the science fiction literary subculture, and a few episodes were even written by leading authors in that tradition. Like them, it featured a military organization, called Starfleet, dedicated not to war but to peace and exploration. But unlike them, it did not serve as a vehicle for education in real science and technology. However much *Star Trek* may have inspired viewers about spaceflight, it almost never instructed them about its factual basis.

Like the early television series, the original 1979 *Star Trek* film was rather intellectual in quality, and I must admit admiring it greatly. In 2009, the franchise experienced a *reboot* through the more action-oriented film titled simply *Star Trek*, which takes place in an alternate universe where history has taken a different course, and the characters are younger versions of the crew from the original series. To fans of the original conception, which stressed philosophical debates about ethics and the meaning of life, the reboot borders on treason. Yet art like technology moves onward, and the reboot may also have moved *Star Trek* away from contemplating the human future in the universe, to stimulating emotional excitement by means of information technology. In that connection, I watched the 2013 sequel, *Star Trek Into Darkness*, on a home wide-screen, high definition "smart television" in three dimensions.

8.3 Non-series Less-Popular Films

Table 8.3 lists the 33 non-series films among the 80 which fewer than 10,000 Netflix customers rated, beginning with *Frau im Mond*, which we have already discussed. The second one in the table, *Destination Moon*, was largely based on a 1947 juvenile science-fiction novel, *Rocket Ship Galileo*, by Robert A. Heinlein (Heinlein 1947). A challenge for the author was how to tell a story in which young readers can vicariously participate in the first lunar expedition. He achieved this by assuming that flight to the Moon was easier than it proved to be, largely because a radical but plausible propulsion method would be used, as the novel's Wikipedia article explains: "After World War II, three teenage boy rocket experimenters are

Table 8.3 Relatively unpopular spaceflight films

Title	Year	IMDB rating	IMDB raters	Netflix mean	Netflix raters	IMDB rank	Netflix rank
Frau im Mond	1929	7.3	1,056	3.35	119	14.5	42.0
Destination Moon	1950	6.3	1,614	3.18	669	50.5	48.0
Rocketship X-M	1950	4.9	1,096	2.93	282	68.5	64.0
When Worlds Collide	1951	6.7	3,281	3.59	1,852	30.5	22.5
Cat-Women of the Moon	1953	3.3	642	2.54	186	77.0	77.0
Project Moonbase	1953	2.4	531	2.66	134	80.0	72.5
Conquest of Space	1955	6.0	635	2.97	104	55.5	62.0
This Island Earth	1955	5.7	4,515	3.03	387	59.0	58.0
The Phantom Planet	1961	2.9	1,580	2.60	99	79.0	74.0
First Men in the Moon	1964	6.6	2,075	3.36	872	35.0	41.0
Barbarella	1968	5.8	14,524	2.99	5,922	57.5	60.0
Marooned	1969	5.6	2,176	3.07	681	60.0	56.0
Silent Running	1972	6.7	11,329	3.17	4,281	30.5	50.0
Dark Star	1974	6.5	10,617	2.83	1,730	40.0	67.0
Flesh Gordon	1974	4.3	2,336	2.59	898	74.0	75.0
Capricorn One	1978	6.8	9,858	3.08	1,723	27.0	54.0
Battlestar Galactica	1978	6.4	3,700	3.33	2,675	46.0	43.5
Meteor	1979	4.8	3,565	2.74	997	71.5	70.0
The Black Hole	1979	5.8	11,221	3.21	6,308	57.5	47.0
Galaxina	1980	3.0	1,328	2.03	316	78.0	80.0
Battle Beyond the Stars	1980	5.2	3,312	2.89	655	65.5	66.0
Saturn 3	1980	4.7	3,713	2.45	1,842	73.0	79.0
Outland	1981	6.5	12,613	3.41	5,192	40.0	38.0
Spacehunter	1983	4.9	2,486	2.76	728	68.5	69.0
The Ice Pirates	1984	5.3	4,770	3.43	3,025	63.5	35.5
Explorers	1985	6.4	7,879	3.80	5,025	46.0	11.5
Space Camp	1986	5.4	6,518	3.42	6,571	62.0	37.0
Space Truckers	1996	4.9	3,855	2.73	381	68.5	71.0
Armageddon	1998	6.4	181,911	3.93	8,552	46.0	8.0
Wing Commander	1999	3.8	10,861	2.79	4,981	75.0	68.0
The American Astronaut	2001	7.2	1,710	3.07	473	19.5	56.0
Ghosts of Mars	2001	4.8	27,874	2.66	8,517	71.5	72.5
The Adventures of Pluto Nash	2002	3.6	13,431	2.59	5,393	76.0	76.0

recruited by one boy's uncle, Dr. Cargraves, a Nobel Prize-winning physicist who had worked on the Manhattan Project, to refit a conventionally powered surplus 'mail rocket.' It is to be converted to run on a thorium nuclear pile which boils zinc as a propellant". When I read the novel as a child, this seemed quite plausible since I had some powdered zinc in my chemistry set, and could create a powerful flash by igniting it.

Destination Moon begins with real footage of one of the many American test launches of a captured German V-2 rocket, at White Sands proving ground in New Mexico, probably in 1946. The view switches back and forth from the real launch to actors watching it from a bunker, as if they were controlling it. The launch film was heavily scratched, apparently taken from a well-worn newsreel, but in the story the rocket was supposed to crash, so the last few frames of the rocket are of a model hitting the ground and exploding, but with scratches added to the film to make it seem part of the real footage. This scene is an odd kludge, perhaps necessitated by the low budget of the movie, but making a nice connection between real rockets and fictional ones. The narrative makes the scene a test failure of a new rocket designed to launch the first Earth satellite.

Watching in horror are two of the four main characters in the film, Dr. Charles Cargraves who led the development effort, apparently this time without help from his nephew, and his friend General Thayer who is also a spaceflight enthusiast. They discuss the possibility that the failure might have been sabotage, and share fears about whether their military sponsors will continue to invest in space technology after this disappointment. The action resumes 2 years later, as Thayer convinces aerospace industrialist, Jim Barnes, to recruit other industrialists to fund an expedition to the Moon. Barnes explains that they are ready to build a Moon rocket, if sufficient investments can be found. The industrialists respond positively to an idealistic appeal, but their commitment is clinched when Thayer informs them that other nations are working on the same thing, and the first one to get to the Moon can use it as an missile base from which to dominate the Earth. The planned ship is shaped somewhat like a V-2, and is inspired by the general design proposed in a popular and heavily illustrated 1949 book, *The Conquest of Space* by artist Chesley Bonestell and Willy Ley (Bonestell and Ley 1949). The film credits Bonestell as its "Technical Advisor of Astronomical Art."

The rocket would have an atomic energy engine, generating 3,000,000 pounds of thrust, and an exhaust velocity for its jet of 30,000 feet per second. This is achieved not by vaporizing zinc but heating ordinary water into high-temperature steam by running it through a nuclear reactor. By the early 1940s, scientists in the German V-2 program, notably Walter Thiel and Krafft Ehricke, had conceived of this general approach, but imagined using hydrogen as the reaction mass, stored in liquid form. Heinlein was apparently aware of this, and Willy Ley, who had close ties with both Heinlein and Bonestell, certainly was (Kingsbury 1975; Ley 1951). When the ship is nearly ready for testing, a government agency blocks further progress, not because of its own safety concerns, but because the general public has begun to express opposition. One shot in the film shows a newspaper headline,

"Mass Meeting Protests Radioactive Rocket." Cargraves, Thayer and Barnes decide to launch immediately, before the authorities can prevent them.

Much of the film is a dramatized introductory course on spaceflight. Barnes shows his fellow industrialists a specially prepared Woody Woodpecker cartoon that explains how rockets function, and how they can propel a craft in airless space. During the launch of the realistic nuclear manned vehicle, the force of acceleration realistically presses the crew back into their couches. Once the ship reaches sufficient velocity to reach the Moon, the engine cuts out, and the crew experiences weightlessness. On the way, they are forced to go outside in their spacesuits to repair an antenna, providing another lesson about the principles of astronautics. Excessive fuel is used in the landing, so there is not quite enough for the return journey, but the scientists solve this problem by discarding all unnecessary equipment so their spaceship will be light enough for the remaining fuel.

I first saw this film at the age of nine during its original release, and as best I can recall I understood what it was trying to teach me, and even 2 years earlier I had seen a lunar eclipse and learned the names of the planets. Thus for many young people of that era, *Destination Moon* either instructed or confirmed their understanding that spaceflight was technically possible, and could deliver human beings to radically unfamiliar environments. The Moon shown in this film had more irregular terrain than the real one, but in 1950 nobody knew the surface details. Within the science fiction community, it was indeed assumed that spaceships would be nuclear powered.

The producer of *Destination Moon*, George Pal, was one of the most innovative people in Hollywood, and he proceeded to produce three more space movies, including a marvelous version of *War of the Worlds* in 1953, which does not depict human spaceflight and thus will not be covered here. In 1951, he produced *When Worlds Collide*, based on the novel of the same name by Edwin Balmer and Philip Wylie (Balmer and Wylie 1933). The Earth is about to be destroyed by a giant rogue planet that recently entered the solar system after millions of years of roaming the galaxy, but its companion planet is more earthlike and may enter a stable orbit and may become a replacement home for humans if only a few of them can escape to it. Much of the movie concerns the struggle to build a spaceship, but the most interesting part is the brief launch scene. The spaceship begins sitting horizontally on the equivalent of a rocket-propelled railway car. At launch, both accelerate together. The track curves somewhat downward, then up a nearby mountain. When the ship reaches the end of the track it is going nearly vertically, and very fast, boosted by the launch device which now drops away. Jet Assisted Takeoff (JATO) had already been developed in reality for ordinary aircraft, but this launch system seems derived from one outlined by Eugen Sänger, a space rocket enthusiast who competed for German military funds against the Peenemünde team in the early 1940s.

Inspired by the Bonestell and Ley book, but more directly representing the ideas of Wernher von Braun, George Pal released a film titled *The Conquest of Space* in 1955. An artistic and financial disaster, it communicated reasonably good technical ideas, but had poor story and dramatization. In Table 8.3, we see that Netflix customers gave a mean preference rating of 3.18 to *Destination Moon*, awarded a somewhat higher 3.59 to the less educational *When Worlds Collide*, and then a lower 2.97 to *The Conquest of Space*. Of course the chief popularity burden *Destination Moon* carries today is the fact that people really have been to the Moon, during the Apollo Program, without benefit of nuclear propulsion, and thus with less consequential result than had history emulated the movie.

Prior to the movie *Conquest of Space*, its general view of spaceflight had already been popularized through a high-visibility series of feature articles in *Collier's Magazine*, later expanded into two books, *Across the Space Frontier* and *Conquest of the Moon*, involving both von Braun and Ley (Cornelius 1952; Cornelius 1953). They also adapted von Braun's more technical book *The Mars Project* to produce the popular *The Exploration of Mars* (von Braun 1953; Ley and von Braun 1956).

First, multi-stage rockets would loft materials into Earth orbit for construction of a space station, with plans to retrieve and reuse the launch stages, and a winged top stage capable of landing back on Earth like the Space Shuttle. The second step would be a large-scale, three-ship expedition to the Moon, assembled at the space station and returning the large crew to it. Finally, a vast expedition to Mars, with winged landers that would shed their wings for launch back to the space station. This was all envisioned to be propelled by chemical rockets like those used by the V-2 or the actual Apollo Program, but on a vastly greater scale that would establish a permanent infrastructure in Earth orbit and beyond, as the basis for further exploration. I read the *Collier's* articles when they were published, and over the years struggled against the tendency to critique the real space program as so far less than we had hoped for.

Of the remaining films in Table 8.3, three deserve brief mention. *Marooned* was released in 1969, the year of the first Moon landing, and this rather realistic film focuses on problems with an Apollo module: "Three American astronauts are stranded in space when their retros won't fire. Can they be rescued before their oxygen runs out"? Released in 1972, the year of the last Moon landing, *Silent Running* connects spaceflight with the Environmentalist Movement: "In a future where all flora is extinct on Earth, an astronaut is given orders to destroy the last of Earth's botany, kept in a greenhouse aboard a spacecraft". Writing in the journal *Space Policy*, Kim McQuaid argues that the spaceflight social movement has neglected potential connections to environmentalism, to its detriment, and this film is probably the best illustration of their potential harmonies (McQuaid 2010). Dating from 2 years later, *Dark Star* is either a "dark" comedy, or a brooding philosophical masterpiece about the futility of not only spaceflight but of human life in general: "Low-budget story of four astronauts in deep space, whose mission is to destroy unstable planets in star systems which are to be colonized. The late Commander Powell is stored in deep freeze, where he is still able to offer advice. As their mission nears completion, they must cope with a runaway alien which resembles a beach-ball, faulty computer systems, and a 'smart bomb' who thinks it is God".

8.4 Five Dimensions of Spaceflight Movies

Immense movie-preference datasets like those collected by Netflix and IMDB can be analyzed in all the ways that have proven useful for traditional questionnaire data. Here we shall illustrate just one, the application of factor analysis to identify major dimensions of meaning that shape preferences. This requires us to focus on movies that were rated by relatively large numbers of respondents, not only to strengthen the statistical reliability of the results, but simply to ensure that all possible pairs of films were rated by significant numbers. Factor analysis begins with a matrix of correlations between pairs of items, so it is necessary to find a balance between a large total number of respondents, and large fractions rating each pair. That has the effect of focusing on people who rated many films, although not all of them, who also tend to be experts on the genre and thus well able to judge its subgenres. Our sample of films will include the first one in each of the popular *2001*, *Star Wars*, *Star Trek*, *Alien*, and *Riddick* series, plus the remake of *Solaris* and the 18 other popular films listed in Table 8.4, all of which were rated by more than 10,000 respondents.

Table 8.4 Highly popular spaceflight films

Title	Year	IMDB rating	IMDB raters	Netflix mean	Netflix raters	IMDB rank	Netflix rank
The Right Stuff	1983	7.9	29,916	3.99	33,906	8.0	7.0
Dune	1984	6.5	58,499	3.49	17,868	40.0	32.0
The Last Starfighter	1984	6.5	16,816	3.60	28,678	40.0	19.5
Enemy Mine	1985	6.7	18,917	3.59	16,025	30.5	22.5
Spaceballs	1987	7.0	78,248	3.54	75,801	24.0	28.5
Total Recall	1990	7.5	124,815	3.61	70,176	11.0	17.5
Stargate	1994	6.9	79,803	3.86	28,557	25.5	9.0
Apollo 13	1995	7.5	123,821	4.08	98,116	11.0	6.0
Event Horizon	1997	6.6	65,229	3.17	13,482	35.0	50.0
Starship Troopers	1997	7.2	131,769	3.37	25,011	19.5	40.0
Contact	1997	7.3	111,753	3.70	36,991	14.5	15.0
Lost in Space	1998	4.9	40,641	2.92	26,457	68.5	65.0
Deep Impact	1998	6.0	82,142	3.33	32,361	55.5	43.5
Galaxy Quest	1999	7.2	74,931	3.54	36,824	19.5	28.5
Red Planet	2000	5.5	30,588	3.07	20,445	61.0	56.0
Mission to Mars	2000	5.3	41,414	3.00	34,079	63.5	59.0
Space Cowboys	2000	6.3	38,951	3.38	46,533	50.5	39.0
The Htchhiker's Guide to the Galaxy	2005	6.7	91,667	2.98	28,454	30.5	61.0

The earliest film in the table was the semi-documentary 1983 drama, *The Right Stuff*, about the Mercury astronauts. The most recent is the 2005 satire, *The Hitchhiker's Guide to the Galaxy*, described as follows in its IMDb synopsis:

> Everyone has bad mornings. You wake up late, you stub your toe, you burn the toast... but for a man named Arthur Dent, this goes far beyond a bad day. When he learns that a friend of his is actually an alien with advanced knowledge of Earth's impending destruction, he is transported off the Earth seconds before it is exploded to make way for a new hyperspace motorway. And as if that's not enough, throw in being wanted by the police, Earth II, an insane electronic encyclopedia, no tea whatsoever, a chronically depressed robot and the search for the meaning of life, and you've got the greatest adventure off Earth.

As a "search for the meaning of life," or for that matter the meaning of spaceflight, this paroxysm of sarcasm could hardly be more different from *The Right Stuff*, which depicts spaceflight in sincerely heroic terms, as well as with technical accuracy. It would be too glib to deduce that faith in spaceflight had crashed between 1983 and 2005, and the other films in Table 8.4 show great variety. But this gloomy assessment may contain a degree of truth.

A total of 227,275 respondents in the Netflix dataset rated at least one of the 24 movies used in the factor analysis. However, 78,475 of them rated only one, so correlations cannot be calculated for them. In order to achieve that balance between reliability afforded by large numbers, and the expertise provided by respondents who had rated many films, after some explorations I settled on a subset of 1,832 respondents who had rated at least 21 of the 24. Expert users of factor analysis tend to prefer "listwise deletion," based on only those respondents who rated all the items included in the analysis. However, that number was only 138 in this case. The alternative I used was "pairwise deletion," which calculates the correlation between two films based on all the respondents who rated both of them. Table 8.5 reports the results of an exploratory analysis that identified five factors, listing all loadings of 0.40 or greater.

Factor 1 reflects liking sci-fi movies of the 1998–2000 period, with two Mars voyage movies heading the factor. The Netflix data were part of the 2006 release of data up through 2005, and these were most of the well-known recent films other than sequels. The earliest film in the factor, *Stargate*, had been transformed into a television series, so it was more recent in people's minds than its 1994 release date would suggest. It is not uncommon for the first factor in analysis of preference data to represent a very general dimension, with less conceptual specificity than later factors. I interpret this one simply to represent liking recent sci-fi films, or indeed liking sci-fi more generally.

Factor 2 represents emotionally intense movies focused on superheroes, blending action with depth psychology. It may seem surprising to see *Star Wars* and *Star Trek* in the same factor, given that they represent different traditions within science fiction, but both emphasize the personalities of the main characters.

Factor 3 collects together three comedy movies, *Galaxy Quest*, *Spaceballs*, and *Hitchhiker's Guide*, with two rather surreal early films that have some satirical

Table 8.5 A Factor analysis of popular spaceflight films

Spaceflight Movie	Year	Factor 1	Factor 2	Factor 3	Factor 4	Factor 5
Mission to Mars	2000	0.79				
Red Planet	2000	0.76				
Lost in Space	1998	0.66				
Deep Impact	1998	0.63				
Event Horizon	1997	0.58				
Space Cowboys	2000	0.54				
Pitch Black	2000	0.49				
Stargate	1994	0.47				
Starship Troopers	1997	0.46				
Alien	1979		0.56			
Star Wars: A New Hope	1977		0.55			
Dune	1984		0.48			
Total Recall	1990		0.47			
Star Trek: The Motion Picture	1979		0.43			
Galaxy Quest	1999			0.71		
The Last Starfighter	1984			0.52		
Spaceballs	1987			0.51		
Enemy Mine	1985			0.50		
The Hitchhiker's Guide to the Galaxy	2005			0.49		0.46
Apollo 13	1995				0.73	
The Right Stuff	1983				0.62	
Contact	1997				0.47	
Solaris	2002					0.71
2001: A Space Odyssey	1968					0.65

or quirky qualities. IMDb's describes *The Last Starfighter* as a marriage between fantasy and reality: "A video-gaming boy, seemingly doomed to stay at his trailer park home all his life, finds himself recruited as a gunner for an alien defense force". *In Enemy Mine*, a human and an alien find themselves marooned together amidst a galactic war: "In the end the human finds himself caring for his enemy in a completely unexpected way" (http://www.imdb.com/title/tt0089092/).

Factor 4 is headed by two factual historical movies dramatizing real spaceflight history somewhat accurately, *Apollo 13* about the accident that affected one of the expeditions to the Moon, and *The Right Stuff*, about the Mercury Project. *Contact* is more visionary, while having sound grounding in real science, as the IMDb synopsis explains:

> Astronomer Dr. Ellie Arroway has long been interested in contact to faraway lands, a love fostered in her childhood by her father, Ted Arroway, who passed away when she was 9 years old leaving her then orphaned. Her current work in monitoring for extraterrestrial life is based on that love and is in part an homage to her father. Ever since funding

from the National Science Foundation (NSF) was pulled on her work, which is referred to some, including her NSF superior David Drumlin, as more science fiction than science, Ellie, with a few of her rogue scientist colleagues, have looked for funding from where ever they could get it to continue their work. When Ellie and her colleagues hear chatter originating from the vicinity of the star Vega, Ellie feels vindicated. But that vindication is short lived when others, including politicians, the military, religious leaders and other scientists such as Drumlin, try to take over her work. When the messages received from space are decoded, the project takes on a whole new dimension, which strengthens for Ellie the quest for the truth. Thrown into the mix are the unknown person who has up until now funded most of Ellie's work and what his motivations are, and Palmer Joss, a renowned author and theologian, who despite their fundamental differences in outlook, is mutually attracted to Ellie, that attraction based in part on intellect and their common goal of wanting to know the truth.

Note the emphasis on seeking the truth, in this synopsis, which is the goal of every true scientist. I also could not help but notice the reference to my employer, the National Science Foundation, and the issue of whether public agencies like it are capable of supporting transformative research.

Truth is obscure in the films making up Factor 5, but transformation of consciousness is vivid. *Solaris* and *2001* are two mystical movies about the meaning of existence, with super-intelligences in the background. Apparently, the respondents see this quality also in *The Hitchhiker's Guide to the Galaxy*, the only film that was placed in two factors by the criterion of at least 0.40 loadings.

Factors 2 through 5 do seem to identify four different meanings that spaceflight may have for movie viewers: (1) an environment in which richly-described human personalities can experience life, (2) an exotic setting for comic surrealism, (3) an expression of heroic possibilities in science and technology, and (4) philosophical contemplation of the mysteries of existence. The average Netflix rating across all respondents for the nine movies in Factor 1 is 3.30, which can be taken as the norm for popular sci-fi movies. The average for Factor 2 is much higher, 3.83, and that for Factor 3 is between them at 3.45. This suggests that strong characters or comic surrealism can add meaning to a sci-fi movie, but that characterization is more powerful than satire. The highest mean, 3.92, is for the three realistic films in Factor 4, and the lowest, 2.98, is for the philosophical triad in Factor 5. These express two different sophisticated meanings spaceflight can have, so the fact that realism trumps philosophy is an interesting finding.

8.5 A Content Analysis

Given enough time, inspiration, and support, a sci-fi franchise could maximize all four of the conceptual dimensions discovered in the factor analysis of the two dozen separate movies. A particularly rich example is the vast corpus of science fiction drama and literature called *Babylon 5* and created by J. Michael Straczynski, a remarkably creative writer whose education included a double college major in psychology and sociology, and who made extremely creative use of many new computer technologies (Bassom 1997).

Babylon 5 (B5) was a science fiction television series of the 1990s, whose 5 years of episodes centered on an orbiting space city (Killick 1997, 1998a, b, c, 1999). *Star Trek* fans may suggest it was a copy of their own franchise's *Deep Space Nine* series, but a better case can also be made that *Star Trek* "borrowed" the idea from Straczynski who had showed his ideas to various people in the TV industry before contracting with Warner Brothers in 1991 to do *Babylon 5*. The voiceover introduction to the first year of episodes suggests correctly that technology was incidental to stories of social conflict and philosophical depth:

> It was the dawn of the third age of mankind, 10 years after the Earth/Minbari war. The Babylon Project was a dream given form. Its goal, to prevent another war by creating a place where humans and aliens could work out their differences peacefully. It's a port of call—home away from home for diplomats, hustlers, entrepreneurs, and wanderers. Humans and aliens wrapped in two million, five hundred thousand tons of spinning metal, all alone in the night. It can be a dangerous place, but it's our last best hope for peace. This is the story of the last of the Babylon stations. The year is 2258. The name of the place is Babylon 5 (http://www.midwinter.com/lurk/universe/setting-1.html).

The vast B5 corpus was used as the source for what amounted to creation of an ethnographic questionnaire, but with a rather different methodological conception from that governing conventional survey research. The goal was not to assemble a moderate number of items to be rated by a large number of respondents, but what I call a *massive questionnaire for personality capture*, consisting of a very large number of items to be administered to a single individual. All of the original television programs (120 h of television, not including episodes of the spin-off series, *Crusade*) and books (up to but not including the *Technomage* trilogy) were scrutinized carefully for clear expressions of general ideas about human life and the universe.

Each statement derived from this data collection effort had to be abstractable from the concrete situation in which it appeared. Some are verbatim quotations, but many are paraphrases. Often there was a grammatical shift from a question or command into a statement. The specific identity of the person the character was speaking to or about often had to be generalized. In a few cases one character began a thought and another completed it. Again, the aim was to derive from the full corpus of *Babylon 5* television and literature, a large number of statements expressing views of life and of human beings.

Exactly 10 statements were derived from each single episode of *Babylon 5*, and 20 from each 2-h television movie. Printed works varied in length and the density of concepts, so a maximum of 25 statements were derived from any one book, and 5 from any article or short story, but often fewer. In part, this was done to respect the copyrights on the original material and stay within the limits of fair scholarly use. (Again, most statements were paraphrases rather than exact quotes). However, this was also an attempt to achieve a broad sampling across the varied corpus of material. In the end of this process, a total of 2,000 statements became the heart of a software module called "Wisdom."

This module was part of the research for my recent book, *Personality Capture and Emulation*, and the appendix of that book provides information on how readers can download free copies for their own private use or for serious research projects (Bainbridge 2014b). The questionnaire actually consists of 4,000 questions, because the respondent rates each of the 2,000 items twice. One 1–8 rating scale concerns how false versus true the statement is, in the judgment of the respondent, and the other is a 1–8 scale from unimportant to important. The 2,000 statements cover a vast intellectual and human territory, indeed of extraterrestrial territory as well because six of the recurring characters belong to alien species. Table 8.6 is a summary of the dramatic sources of the statements, tabulating how many of the 2,000 each of the main characters contributed.

Of course the descriptions of the characters in Table 8.6 are not attitudes toward spaceflight itself, but they suggest a range of personalities that might come to value spaceflight, each in its own way. In an earlier essay, I suggested that

Table 8.6 Wisdom module popular culture items drawn from *Babylon 5*

Character	Description of character	Items	Mean true	Mean important
Humans:				
Jeffrey Sinclair	First commander of B5 and representative of Earth Alliance. Perhaps a mild inferiority complex, because as a fighter pilot he feel out of place in the world of diplomacy. Prone to take risks but conceals his feelings behind a mask of formality	86	4.49	4.85
John Sheridan	Second commander of B5 and representative of Earth Alliance. A cheerful war hero with a naturally balanced personality, he endures tremendous stress from those who hate him, or want him to become a messiah	190	4.90	5.05
Michael Garibaldi	Security chief of B5 but erratic, semi-reformed alcoholic, plagued by self doubts and paranoia. His unstable personality makes him work all the harder to preserve security	100	4.62	4.88
Susan Ivanova	Second in command of B5; pessimistic, quirky, with a wry sense of humor; buries herself in her work. Resentful because her mother committed suicide rather than accept drug treatments mandated to control latent telepaths	72	4.82	4.93
Dr. Stephen Franklin	Chief medical officer of B5, but more of a biologist than a doctor, who loves to explore new environments and resists being controlled by bureaucracy. Follows his own personal ethical code	68	4.85	5.12
Marcus	A mysterious ranger, whose sarcasm conceals sorrow, he is capable of both ferocity and determination	36	4.64	4.97

(continued)

Table 8.6 (continued)

Character	Description of character	Items	Mean true	Mean important
Bester	A ruthless leader of the Psi Corps, who pursue rogue telepaths and function as a secret security arm of the Earth government, while seeking to manipulate power for their own gain	72	4.61	4.69
Lyta Alexander	The first Psi Corps telepath assigned to B5, her life and mentality were forever altered by reading the thoughts of Kosh the Vorlon, which caused her to go rogue and become hunted by her former agency	34	4.50	5.32
Byron	A leader of a strange group of independent telepaths, perhaps a messiah or martyr, who views ordinary "mundane" people as inferior	30	4.60	5.37
Extraterrestrials:				
Delenn	Ambassador from the Minbari who recently fought a war with Earth, she is fascinated by humans yet deeply influenced by Minbari religion. She manages this deep contradiction by gradually coming to believe she may be a messiah	147	4.44	5.08
Lennier	Aide to the Minbari ambassador. Priestly, having just left a monastery, worships his ambassador, innocent, submissive, awkward	32	4.22	4.69
G'Kar	The Narn ambassador, fundamentally idealistic yet ambitious. Seeking spiritual enlightenment in ancient scriptures, he often gains unusual insights into the very non-spiritual conflicts raging around B5, through clear-eyed observation	131	4.60	5.08
Londo Mollari	The Centauri ambassador, a decadent aristocrat who mourns lost glories, drinks heavily, and has a sarcastic humor. Open to secret deals and deceptions, if he thinks they will advance his status	147	4.64	5.01
Vir Cotto	Aide to the Centauri ambassador. Timid, lacking self-confidence, yet fundamentally honest. His moral instincts struggle against his sense of powerlessness	37	4.62	4.76
Kosh	The Ambassador from the Vorlons, an extremely enigmatic alien, who wears ponderous and obscuring armor. The methane-breathing species seeks absolute control, in a galactic contest with the Shadows, who seek chaos	23	5.09	5.48

(continued)

Table 8.6 (continued)

Character	Description of character	Items	Mean true	Mean important
Miscellaneous:				
Minor Characters	Run of the mill characters, including some who appear only in a single episode; a diverse group in which some have undefined personalities	636	4.63	5.04
Anonymous	Narrators or unnamed characters that generally express an impersonal viewpoint	159	5.01	5.33

stereotypes of extraterrestrials tend to reflect exaggerated forms of standard human personality types, even in some cases specific psychopathologies (Bainbridge 2014c). The last two columns in the table report my own mean ratings of the statements contributed by each character, which suggest how I think various personalities connect to my of my own values, including pro-spaceflight attitudes. The reader can download the Wisdom software, rate the statements, and immediately gain an analysis from the reader's own viewpoint.

The first two alien characters in the list, Delenn and Lennier, are from a haughty species that once tried to exterminate humans, and I tended to rate their statements lowest on the "true" scale. One of the statements spoken by the character Delenn well communicates the central principle of their caste-ridden society: "Understanding is not required, only obedience." At the opposite end of the list, I give the highest "true" rating to statements by Kosh, the enigmatic Vorlon. Kosh is famous throughout the science fiction subculture for making inscrutable pronouncements hinting at profound wisdom, such as: "The avalanche has already started; it is too late for the pebbles to vote."

I also give somewhat high "true" ratings to anonymous statements and those from the second commander of Babylon 5, John Sheridan. The "anonymous" category consists of statements from television characters who are so minor they lack names and from authors writing about B5 without taking the voices of characters, so in a sense these statements lack personality. Sheridan, the central character of the series, is a Christ-like figure who dies but is reborn. His statements express both optimism and stoicism: "If you're falling off a cliff, you might as well try to fly." "The way to deal with pain is to turn it into something positive."

8.6 Media and Meaning

Science fiction was transformed when it became a popular genre of movies and television, and it is not entirely clear what the consequences were for public appreciation of spaceflight. In the 1912 novel *A Princess of Mars* by Edgar Rice Burroughs, John Carter is teleported to the red planet by a poisonous gas that detached his spirit from his body and allowed him to leap through space, and his return was initiated months later by asphyxiation. The highly unsuccessful motion picture version *John Carter*, released a century later, replaced this supernatural mechanism by mysterious amulets belonging to the Holy Therns that were teleportation devices, leaving to the audience the question of whether they represented advanced technology or the magic of an exotic religion. Neither version of the story employs chemical-fuel rockets, the novel dating from before that means had been conceptualized, and the movie dating from after that technology had exhausted its limited potential.

Star Wars places all its action "long ago, in a galaxy far, far away," and most of the action in *Star Trek* takes place outside our own solar system. Thus they do not need to deal with the fact that the worlds within actual reach of Earth are naturally unsuited for human habitation. *John Carter* begins with a brief image of Mars from a distance, and the voice of a Martian named Tars Tarkas speaks: "Mars. So you name it and think that you know it. The red planet, no air, no life. But you do not know Mars, for its true name is Barsoom. And it is not airless, nor is it dead, but it is dying." Perhaps, outside fantasy mass media, the dream of visiting Mars is also dying. The spectacularly successful 2013 science fiction film, *Gravity*, expresses the gravity of this situation by depicting the catastrophic end to human presence in space, ending the instant the heroine survives the fall back to Mother Earth.

Readers of science fiction literature must absorb each word in the order specified by the author, but they have some freedom to visualize each scene according to their own imagination. Audiences of movies and television are more passive, seeing only what the director and special effects technician provided. The newest science fiction medium, virtual worlds experienced through computers and videogame systems, provide more freedom than any of the traditional media. They allow the user to decide to a significant degree what will happen from moment to moment, and they depict dynamic environments. Thus, they seem more real, more personal, and may therefore have greater influence, notably in helping the user visualize the experience and possible goals of spaceflight.

References

Agel, J. (1970). *The making of Kubrick's 2001*. New American Library.

Bainbridge, W. S. (2011). *The virtual future* (pp. 167–182). London: Springer.

Bainbridge, W. S. (2014a). *Personality capture and emulation* (p. 81). London: Springer.

Bainbridge, W. S. (2014b). *Personality capture and emulation* (p. 206). London: Springer.

Bainbridge, W. S. (2014c). Patterns of extraterrestrial culture. In D. A. Vakoch (Ed.), *Extraterrestrial altruism: Evolution and ethics in the cosmos* (pp. 275–294). Heidelberg: Springer.

Balmer, E., & Wylie, P. (1933). *When worlds collide*. New York: Frederick A. Stokes.

Bassom, D. (1997). *Creating Babylon 5*. New York: Del Rey/Ballantine.

Bonestell, C., & Ley, W. (1949). *The conquest of space*. New York: Viking.

Cornelius, R. (Ed.). (1952). *Across the space frontier*. New York: Viking Press.
Cornelius, R. (Ed.). (1953). *Conquest of the Moon*. New York: Viking Press.
Heinlein, R. A. (1947). *Rocket ship Galileo*. New York: Scribner.
Heinlein, R. A. (1948). *Space cadet*. New York: Scribner.
Killick, J. (1997). *Babylon 5: Signs and portents*. New York: Del Rey/Ballantine.
Killick, J. (1998a). *Babylon 5: The coming of Shadows*. New York: Del Rey/Ballantine.
Killick, J. (1998b). *Babylon 5: Point of no return*. New York: Del Rey/Ballantine.
Killick, J. (1998c). *Babylon 5: No surrender, no retreat*. New York: Del Rey/Ballantine.
Killick, J. (1999). *Babylon 5: The wheel of fire*. New York: Del Rey/Ballantine.
Kingsbury, D. (1975). Atomic rockets. *Analog*, December, pp. 1–17.
Ley, W. (1951). *Rockets, missiles, and space travel*. New York: Viking.
Ley, W., & von Braun, W. (1956). *The exploration of Mars*. New York: Viking Press.
McQuaid, K. (2010). Earthly environmentalism and the space exploration movement, 1960–1990: A study in irresolution. *Space Policy, 26*, 163–173.
Nowlan, P. N. (2010). *Armageddon 2419 A.D. and other works*. Houston, Texas: Halcyon Press.
Verne, J. (1869). *From the earth to the moon*. Newark, New Jersey: Newark.
von Braun, W. (1953). *The Mars project*. Urbana, Illinois: University of Illinois Press.
Weinstein, D. (2004). *The forgotten network: Dumont and the birth of American television*. Philadelphia: Temple University Press.
Wells, H. G. (1901). *The first men in the moon*. London: G. Newnes.

Chapter 9
Simulation

A very new approach to learning what spaceflight means is to explore computer simulations, many of which are marketed as games, and observe how they depict outer space. In 1956, I was one of the proud owners of a Geniac, an inexpensive but sophisticated toy computer developed by Edmund C. Berkeley, a founder of the Association for Computing Machinery (ACM), the central professional organization today in the field of computer science. Geniac was a system of rotary switches, with flashlight bulbs as the output, and rewiring as the programming method. Among the simulations it could produce was a "machine for a space ship's airlock" and a game called "The Uranium Shipment and the Space Pirates" (Garfield 1955). Already by the mid-1970s, the U. S. Space and Rocket Center in Huntsville, Alabama, had among its displays a very simple videogame in which visitors to this museum could try landing the Apollo lunar lander, learning something about how a rocket can decelerate if it is controlled in exactly the right way.

It is hard to say which was the world's first significant computer game, but *Spacewar!* dating from 1962 is often credited with being the landmark that really launched the genre. *Spacewar!* used a video display and simulated with some degree of realism how a rocket in a star's gravity field could be moved from orbit to orbit, as its Wikipedia page explains: "The basic gameplay of Spacewar involves two armed spaceships called 'the needle' and 'the wedge' attempting to shoot one another while maneuvering in the gravity well of a star. The ships fire missiles that are unaffected by gravity (due to a lack of processing time). Each ship has a limited number of missiles and a limited supply of fuel. Each player controls one of the ships, and must attempt to simultaneously shoot at the other ship and avoid colliding with the star".

This chapter will begin by considering two very different kinds of educational simulation, the planetarium program *Starry Night*, and a strategy game depicting the real process of organizing a space program, *Buzz Aldrin's Race into Space*.

W.S. Bainbridge, *The Meaning and Value of Spaceflight*,
Space and Society, DOI 10.1007/978-3-319-07878-6_9

Since our goal is to understand the popular conceptions of spaceflight, and online role-playing games often depict living within a community of other players on alien planets, we will examine six of these virtual universes that illustrate different principles. First we visit a fantasy moon and a fantasy planet in two of the most popular massively multi-player online (MMO) games, *EverQuest* and *World of Warcraft*. A transition from fantasy to science fiction is accomplished in *Star Wars Galaxies* and *Star Wars*: The Old Republic. Finally, two colonization simulations, *EVE Online* and *Entropia*, give us data comparable to those gathered by questionnaires, in the expressed goals of hundreds of player groups.

9.1 Educational Simulations

A very large number and variety of educational computer games either focus on astronomy or use outer space as the context to teach other topics, such as environmental science. An entire book could be written about this complex topic, so here we shall briefly consider two classic examples to establish the outlines of this genre. A good example is the educational astronomy software, *Starry Night*, which is a personal, computer-based planetarium.

When *Starry Night* first launches, it requests the user's location, to be able to customize the display, which in my case specifies the nearby US Naval Observatory. At subsequent launches, it requests permission to check for updates to data such as "asteroid, comet, and satellite positions." Then it asks if the user wants to look at specific timely events. For example, in mid-afternoon on January 1, 2014, it listed: Europa Shadow Transit, Europa Transit, Io Eclipse, Io Occultation, New Moon, and Pluto Conjunction. The first four involve two of Jupiter's moons, and are common prominent events. The New Moon and Pluto Conjunction mean that both objects are very near the direction of the Sun, as seen from Earth. Whether ironically or significantly, googling "Pluto conjunction" turns up a vast number of astrological websites, rather than astronomical ones. Selecting the Europa Transit in *Starry Night* gives a rather close-up view of the planet Jupiter, with its moon Europa crossing the planet from Earth's perspective, closely preceded by its shadow on the planet's surface. The interface allows zooming in or out, speeding up or reversing time, and seeing information about Europa displayed either graphically or as text.

At this point in time, Jupiter was below the horizon, and its moon Io was behind the planet, so Io was doubly eclipsed. Yet the user can zoom around and look at things that would not be visible in the actual sky for the user's home location. In "spaceship mode," I decided to check out Io, by way of a long detour to the vicinity of Saturn, where I experienced a solar eclipse by flying over the dark side of Saturn's largest moon, Titan. Then, I zoomed toward Jupiter, and once I was close, I turned toward Io. When I got there, it was still possible to see both the Sun and the Earth, since Io moves so fast in its relatively tight orbit, and the eclipse would not happen until tomorrow. Users clearly can learn a lot about the structure and dynamics of the solar system, although amateur astronomers would use it frequently for selecting deep-sky objects to observe with their telescopes, rather than the solar system. The frankly enthusiastic text at the beginning of the instruction manual speaks the truth:

> The invention of desktop astronomy software has been the most exciting new development in the astronomy hobby in years, perhaps since that night four centuries ago when Galileo pointed the newly invented telescope at the heavens for the first time! Starry Night is the premier astronomy software package on the market, putting more power and knowledge in your hands than even the world's pre-eminent astronomers had just a few years ago. You can see how the sky will look tonight, tomorrow, or far into the past or future. You can view the stars as they appear from your own backyard, from a country on the other side of the world, or from another planet. You can witness a total eclipse from the Moon, watch the Sun set from the surface of Mars, or even ride a comet. You are limited only by your curiosity (Anonymous 2009).

What the user learns about spaceflight is less clear. The user never sees a ship per se, but movement of the perspective. To its credit, *Starry Night* depicts speed changes as the result of more or less vigorous acceleration, pressing the A key to accelerate, and Z to decelerate, although without expenditure of fuel. Pressing A then releasing it leaves the ship progressing at whatever velocity it reached during the acceleration when the key was pressed. Many commercial games depict spaceships like cars or airplanes, in which the engine must be kept going to maintain constant speed, and turning it off slows the vehicle to a stop, which is totally unrealistic in outer space. However, the speed of light does not represent a limit in *Starry Night*, the stars in the direction of flight do not exhibit red shift as the speed of light is approached, and constant acceleration force does not result in diminishing increments to speed near the speed of light as would happen with a real spaceship. I found that a velocity 40 times the speed of light was comfortable for a brief tour of the solar system, but I had to be ready to decelerate swiftly when approaching a planet, at something in the range of millions of gravities, which I imagine would have produced nuclear fusion of all my atoms, had it been happening in real life.

Among the most highly regarded early examples of educational spaceflight software is *Buzz Aldrin's Race into Space*. Aldrin, of course, was one of the first two humans to reach the Moon, on Apollo 11, one of the many people inside the space program I have interviewed over the years, and this game simulates the competition in the real historical Space Race which he experienced first hand. The version I examined could be played by one person against the computer, or two people against each other, one representing the American space program, and the other the Russian. The instruction manual explains:

> In *Buzz Aldrin's Race Into Space*, you are placed in command of your country's space program. As Mission Director, you'll purchase and develop space hardware, recruit and assign astronauts, plan and initiate missions into space. The first country to successfully complete a manned lunar landing and return to Earth wins the game. This is a strategy-oriented game that requires short and long-term planning. You'll need to determine what space hardware is needed to complete your objectives. While it is certainly not required, it is suggested that you read some of the historical material on the space race. The American and Soviet strategies are quite insightful (Bronner 1999).

Despite some simplification, the game did attempt to depict with historical accuracy the sequence of decisions faced by each side in the competition, including some of the economic, technical, and logistical details. The reference to *strategy-oriented game* places it within a very general genre of strategy games, of which chess is among the most familiar, and to which the *StarCraft* series of space-oriented commercial games also belongs (Farkas 1998). Like *StarCraft*, this game requires making decisions about how to make investments developing capabilities, taking greater or lesser risks, and requires a more analytical approach from the player, than do the role-playing games described later in this chapter. Thus, while one does learn about the real issues faced by both sides in the historical Space Race, much of the learning for students would concern more abstract issues of logistics and planning that are relevant in most technical and economic projects. While abstract, this is far more realistic than the majority of commercial spaceflight games, which raises the issue of how much realism human beings really want when they contemplate spaceflight.

9.2 Fantasy Planets

Many massively multiplayer online role-playing games (MMOs) set some or all of the action on another planet, but almost universally these environments are very earthlike. Thus they reinforce the "other worlds" public perception of planets which supports the idea that we might some day migrate to live on them. At a first approximation, these games and planets belong to the two rather different traditions that interweave in the literature, fantasy and science fiction. We can begin by considering two highly popular fantasy games, *EverQuest* and *World of Warcraft*, because they illustrate some of the game-design factors that determine the nature of the environments.

EverQuest first launched in 1999, and *World of Warcraft* in 2004, although the latter was the fourth stage in a series that began with a two-player game, *Warcraft*, in 1994, and a sequel to the first, named *EverQuest 2*, launched in 2004, so both have complex histories. Both were greatly influenced by the table-top role-playing game *Dungeons and Dragons*, which dates from the 1970s, and thus in their great popularity both represent a major fantasy tradition (Gygax 1979). Players start each game by creating an avatar in a virtual world that is not the Earth, Norrath in the case of *EverQuest* and Azeroth in the case of *World of Warcraft*. However, these are not conceptualized as planets that physically exist elsewhere in our physical cosmos, but as alternative realities. Some time after the original launch of the game, each gained a satellite world, conceptualized as another astronomical body in the alternative universe, Luclin in the case of *EverQuest*, and Outland (or Draenor) in the case of *World of Warcraft*. I thoroughly explored both by running avatars through the two MMOs, and here will analyze their structure and meaning.

Both Luclin and Outland were added in expansions of the original MMO. When a high-investment virtual world is first opened for people to explore online, it is typically quite large in human terms, containing the subjective equivalent of a hundred or more square miles of virtual territory. A major design decision is the extent to which this territory is divided into separate areas, with narrow gateways through which the avatar must pass, connecting *zones* or *instances* that are managed within both the user's computer and the Internet-connected game company's server as separate sections of memory, even in the latter case sometimes maintained on separate pieces of hardware. Early games were heavily instanced, simply because the user's hardware had very limited memory, and the programming techniques that would allow switching data in and out of memory required a fast central processor, so it was simply not feasible to handle the world as a single unit.

In traditional videogames, this technically-mandated modularity harmonized with good gameplay design principles, as comparison with the traditional outdoor sport baseball illustrates. Nominally, each baseball game is divided into nine innings, with each team at bat for roughly half of each inning. One could imagine having the teams switch places after each batter, but then the afternoon would primarily be taken up by walking in and out of the field more than fifty times. This fine-grained turn-taking worked for chess and for the original *Warcraft* game, but not for baseball and not for videogames that take dozens of hours to finish. Or one could imagine having only one inning, in which one baseball team got to earn all its points first, and the tension in the second half of the inning would entirely focus on whether the other team could surpass its score. The traditional videogame was divided into levels, each of which took from a few minutes to over an hour, ending with a battle with a boss enemy, which provided a nice dramatic climax, allowed the player to quit temporarily after each level, and avoided the need for doing a major change of the game machine's active memory during the level.

Given its earlier vintage, *EverQuest* was highly instanced, each of its more than 375 zones functioning as a separate unit. *World of Warcraft* blended the zones together, so that they represented qualitatively different environments, but the avatar could cross most of the boundaries at many points and without any pause while data were loaded into memory. Within a zone, both MMOs had somewhat different areas and organized the action in terms of quests or missions that connected with the environment and each other in various ways. Naturally, the computer code and the database were also structured in terms of units and subunits, so modularity is the name of the game in MMOs, as in baseball. Indeed, many forms of performance art are modular, because the human mind and experience of life are both modular. Theater dramas have acts and scenes, television series have episodes, novels have chapters, and as the Earth rotates, humans experience days and nights.

In MMOs, some of the largest modules are expansions, such as when *EverQuest's* world, Norrath, gained a moon, Luclin, in 2001. In his book about the team that created *EverQuest*, Robert Marks explains that expansions can have different goals, which we here can conceptualize as different potential goals for colonization of real other worlds. *EverQuest* and *World of Warcraft* have the same strategy for earning income for their companies, charging a moderate amount for the player to buy the game, then adding a monthly subscription cost. This motivates the companies to find ways to commit the player to the game, and many subscribe literally for years. Adding high-level material gives the player's original avatar new challenges and a sense of further progress. The alternative is to add new territory that covers the full range of experience levels, and thus is primarily intended for the player to create a second avatar, ascending the experience ladder again, but with a very different set of challenges and experiences. Marks notes that adding Luclin was the latter type of expansion, providing a new environment possessing zones for all the experience levels, but also challenging the design team because they also used this moon as an opportunity to take a step upward in the quality of the computer graphics (Marks 2003).

Travel from Norrath to one of the action zones on Luclin is slightly arduous, but taking a few minutes rather than the 5 days between Earth and Moon, and having zero economic cost. Still, there is a sense of separation, and Luclin is a complete world. Notably, it possesses a hub zone, called Nexus, which is more-or-less a duplicate of a hub zone connecting the action zones of Norrath, called the Plane of Knowledge. The Nexus is a subterranean city where no enemies exist, but containing a large number and diversity of friendly non-player characters who sell things and assign training missions to the avatar of the player. Like the Plane of Knowledge back on Norrath, it links to a secondary zone called the Bazaar, where players can sell virtual goods to each other, and it is home to intensive informal social life that brings players together, encouraging them to form the enduring player groups called *guilds*. From the Nexus, an avatar can walk through tunnels to a variety of outdoor action zones, most of which do not seem extraterrestrial at all.

The first exception a player typically visits is Marus Seru, gray, dusty hills which feel like the surface of our own Moon but with a substantial atmosphere. Oddly, many alien animals dwell there, despite the lack of water or vegetation. Since the purpose of Luclin was to provide an alternative world for players of all levels to explore, to minimize monotony it needed to have a great variety of environments. At the opposite extreme from Marus Seru was The Twilight Sea, described thus on one of the *EverQuest* wikis:

> The Twilight Sea is probably more land than it is water, consisting of several islands as well as a hidden area in its north accessible only by an underwater tunnel. Katta Castellum sits in the east, its towering walls and guarded boundaries difficult to miss. Numerous city residents can be found on a nearby island containing two fishing docks and an inn called Jern's Rest. The Twilight Sea also connects to the dry, barren part of Luclin: Past the range of cliffs and mountains in the west is the Scarlet Desert.

The Scarlet Desert resembles the version of Mars described in much traditional science fiction, such as *A Princess of Mars* by Edgar Rice Burroughs and *Red Planet* by Robert A. Heinlein (Burroughs 1917; Heinlein 1949). This zone of Luclin is inhabited by intelligent aliens as well as fierce beasts: "While its long-dry riverbeds indicate it wasn't always the case, the Scarlet Desert is one of the driest areas on Luclin (save for a small oasis in its south)". Beyond this hazardous but survivable zone lies the Grey: "You'll want an Enduring Breath effect or item before travelling into the Grey. Its desert wasteland, located on the light side of

Luclin, has been long-exposed to the vacuum of space and is void of oxygen". The game's designers attempted to provide a plausible explanation for why a moon could possibly have such diverse terrain, as summarized on an *EverQuest* wiki:

> Luclin is one of two known moons of the planet Norrath. (The other being Drinal.) Around 3711, Luclin exploded. While the majority of the lunar surface remained in place, the satellite was severely damaged, splitting into multiple pieces. Some of the ejecta from the explosion, including portions of the surface of Luclin and likely portions of the interior of the planetoid, was caught in the gravitational pull of Norrath and struck the surface. While it can be assumed that most of the material that made planetfall did so in the oceans (as Norrath is predominantly water-covered), many meteoroids did strike the surface of Norrath. (Thereby becoming meteorites by definition.)... The moon of Luclin is in perfect balance – one side bathed forever in light, and the other steeped in darkest night. The earliest arrivals felt an eerie and disturbing presence when approaching the Twilight between the two areas. Those not native to the moon quickly learned to avoid these dangerous lands. Ironically, it is said that those of good and honest nature made their homes in the darkness, and those of a darker nature chose the light.

The relevance for real interplanetary colonization is twofold: (1) the extent of separation from our homeworld, and (2) the extent to which the other world is designed for ordinary humans versus for an elite class, perhaps limited to rich and powerful people who can afford the cost of space travel. Luclin is entirely self-contained. While we can imagine some shipment of small manufactured goods from Earth to a colony on the Moon or Mars, such as computer chips and pharmaceuticals, extensive transportation of goods would be uneconomical once a colony was well-established. During the 1950s, comedian Ernie Kovacs included in his radio show a recurring skit, "Space Commuter," which imagined a man living on the Earth would commute every day to work in a department store on the Moon, often forced to take a local spaceship that went by way of Saturn. I remember one episode in which the commuter learned that his robot companion had a bomb hidden in one of its many ears, which required unscrewing all of them and then disposing of the bomb through the razorblade slot in the men's room of the commercial spaceship. This is ridiculous, of course, as it was meant to be. Luclin teaches players that planets and moons are separate and self-contained, although it minimizes the cost of travel between Luclin and Norrath.

Luclin is suitable for beginning avatars, and three of its action zones are survivable for avatars of the very lowest experience levels: Shadeweaver's Thicket, Shadow Heaven, and Shar Vahl. But it also has zones for every range of experience levels up through Vex Thal which is suitable for avatars of experience levels 55 through 70. This raises the question of whether a Lunar or Martian colony would be open to human beings of all degrees of wealth, power, and ability, or only for an elite. Outland, the second planet in *World of Warcraft*, was an expansion intended only for avatars that had already reached at least level 58, and that raised the level cap from 60 to 70 when it opened in 2007 (Lummis and Kern 2007; Bainbridge 2010). As of 2013 when I most recently explored both MMOs, the *EverQuest* experience cap was 100, while that of *World of Warcraft* was 90, so while not identical, their experience ladders are similar.

To survive in Outland, one must be an experienced player and operate a high-level avatar, analogous to the high level of skill and durable temperament of real astronauts, like those in the Mercury Program who possessed the "right stuff." We can speculate how colonists might be selected for a Mars colony, especially after the period of its original establishment. Perhaps a combination of metal and physical tests would be required, comparable to the series of steps required to join a top sports team of an elite Ivy League university. Perhaps as well, very rich people could simply pay the costs of their own emigration. But given the literally astronomical costs, and months spent in transit, even the rich will not commute between planets.

Outland and Azeroth have a major difference from Luclin and Norrath, in that they explicitly reject the standard astronomical concept of what a planet is. *World of Warcraft* substitutes the ancient notion that the world is flat. Azeroth consisted originally of two vast continents, now four of them, separated by an ocean that results not from the gradual continental drift that moved the continents apart on the surface of the Earth, but ripped apart suddenly when Elves misused their magical powers. The ocean can be crossed by ships or zeppelins, but only across the area that was the point from which the continents separated, not through the equivalent of trans-Pacific travel. It is worth noting that in the four novels by Edgar Rice Burroughs about the planet Venus, conceptualized as a steamy jungle called Amtor where the natives never saw the stars and thus lacked astronomy, the natives had an identical view of their world, and first editions included a map of Amtor supposedly drawn by the natives that resembles the map of Azeroth (Burroughs 1946).

Outland is different from Azeroth and well as Luclin, because Outland is not only flat but disintegrating, and especially in Hellfire Peninsula the land consists of a collection of levitating rocks, from small to vast. Despite the fact that often there is little solid material underfoot, gravity is still earthlike and operates in a single "down" direction. This is actually a nice metaphor, because many societies in *World of Warcraft* are also disintegrating, while the most primitive ones have yet to unify fully. As a fit environment for dramatic conflict, the world of *World of Warcraft* is not so much ignoring the laws of nature discovered by science, as depicting the repeal of those laws. Indeed, the word *world* has two meanings in this context. It refers both to the social world and the physical world, and as one disintegrates, so does the other. That is one way also to distinguish fantasy from science fiction. Fantasy makes greater use of poetic metaphors, whereas science fiction accepts more of the constraints imposed by scientific rigor.

9.3 The Transition Between Fantasy and Science Fiction

As noted in Chap. 7, the action-adventure genre represented by the Mars novels by Burroughs serves as a pro-spaceflight transition between fantasy and science fiction. Chapter 8 noted that the more recent and highly popular *Star Wars* franchise is a direct descendant of Burroughs and *Flash Gordon*, thus within that same genre. Many *Star Wars* computer games have been created, and the most extensive current one is the MMO *Star Wars: The Old Republic*, which replaced the earlier and much-respected MMO, *Star Wars Galaxies*. Together, these two MMOs well illustrate the characteristics that the large game-playing segment of the public expects spaceships and planets to have.

The general public often refers to vehicles for interplanetary travel as *spaceships*, yet this term is very seldom used in NASA or the aerospace industry, which prefers more abstract terms like *spacecraft* or *space vehicle*. This usage reflects fundamentally different ways of thinking about space travel, and the popular mind likes the analogy with ocean voyages. NASA's classic history of the Mercury Project was titled *This New Ocean*, in deference to this popular view (Swenson et al. 1966). However, a different popular analogy is with aviation, as reflected in the term *spaceflight* and in the fact that early astronauts were former fighter plane pilots. A third metaphor, less mentioned but probably still shaping popular conceptions, sees a spacecraft as a kind of *residence*, a home very far away from home, a safe place for rest and domestic chores while exploring exotic worlds. These conceptions of space vehicles are very clearly illustrated in the two *Star Wars* MMOs.

Launched in June 2003, for 8 years *Star Wars Galaxies* (SWG) allowed thousands of fans to experience Tatooine, Naboo, and other favorite fictional planets at a time period in the midst of the movie events (McCubbin et al. 2005; McCubbin 2005). *Star Wars*: The Old Republic (SWTOR) was set 3,500 years earlier in galactic history, to facilitate telling new stories unconstrained by events in the movies, yet the level of technological development at that earlier point in history was identical (Searle 2011).

The Millennium Falcon model YT-1300 vehicle in the original 1977 *Star Wars* movie is clearly a ship, although it has two gun turrets like those on bomber aircraft of the Second World War, from which Luke Skywalker and Han Solo can shoot at small enemy ships that operate like fighter aircraft. The Falcon is spacious, containing many rooms, pieces of furniture, and storage compartments. Gravity functions just as it does on a ship at sea, or even more like in an ordinary terrestrial home, because at sea the vessel would constantly pitch and roll, but in the film the Falcon seems totally stable most of the time. Weightlessness as experienced by real space crews is not depicted.

When Luke attacks the Death Star, he pilots a small winged craft that looks very much like a classic jet fighter, perhaps an F-104 that first flew in 1954. Darth Vader's cronies fly small fighters of a very different design, with solar panels on the sides, explicitly called TIE fighters, with TIE standing for Twin Ion Engine. Of course it would really be possible to build a small manned spacecraft propelled by ion engines, using solar panels to produce the needed electric power, but its acceleration would be so gentle as to be imperceptible, and zooming around in a dogfight would be absurd.

In October 2004, the Jump to Lightspeed expansion of *Star Wars Galaxies* added individually owned spacecraft, and over time a number of options were added. At launch, SWG had offered public space transport between planets, but the method did not depict the voyage itself. Rather, a player would go to the nearest spaceport, buy a ticket from a vending machine, present that ticket to a droid standing beside a spaceship, and then the scene would quickly shift to the spaceport on the distant planet. Jump to Lightspeed added the experience of flying through space, and shooting at enemy craft, with a display that primarily depicted the view through the windshield surrounded by many instruments and controls. Wikipedia explained the fully developed system:

> As characters advance in their piloting professions, they gain access to a widening variety of tactics, starship chassis, and starship components. Their ships can be completely customized with components looted from enemies or crafted by shipwrights. Available chassis include the X-Wing and Y-Wing for Rebels, TIE Fighters and TIE Bombers for Imperials, and new Hutt and Black Sun ship designs for Freelancers. Characters who have mastered a piloting profession get access to PoB (Player on Board) ship designs such as the famed YT-1300. PoB ships allow characters to walk around the interiors (which can be decorated just like a building on the ground) and man additional shipboard stations such as laser turrets.

The model of PoB ship I owned in SWG was not intended for combat, but was the Sorosub luxury yacht, which I earned simply by subscribing to the game for 180 consecutive days. It functioned as a mobile home, because it possessed many rooms, connected by stairways and an elevator, and could hold as many as 100 items including furniture. In its very last days, SWG added a feature that caused considerable chaos and fun, and undoubtedly had been withheld for years because of its disruptive nature: flying within the atmosphere of an inhabited planet. The yacht zoomed around the skies of Tatooine, crashing into the mountains and the buildings of towns like Mos Eisley, visible to any players standing below and sometimes obscuring their view.

SWG was very much intended as a virtual world in which players would live, what is called a *sandbox* game, providing relatively few story-based missions but offering great freedom to create one's own missions and virtual goods. Thus, one of my four avatars was an engineer, who mostly made droids but also built a few small spacecraft of the fighter type. Thus was done through elaborate work to garner mineral resources from the environment, use virtual machinery to construct components, and then assemble the pieces into the vehicle. My engineer also built about ten houses, which I placed on frequently visited planets, which functioned as homes and storage places like the Sorosub yacht, but unlike it were not mobile.

Star Wars: *The Old Republic* is a relatively new form of *themepark* game, with very elaborate stories that require completion of complex mission arcs that are punctuated with *cutscenes*, which are like animated, fully-voiced scenes from motion pictures. We do not have a standard term for such games yet, although the early levels of the 2008 MMO *Age of Conan* were also heavy with cutscenes, and *The Secret World* which launched in 2012 is filled with them. They could be called *theatrical* games, not only because of the cutscenes, but also because the player is acting out a pre-scripted story, like some that has long been common in solo-player computer games, but not in MMOs. SWTOR gives the player secondary avatars, called *companions*, to assist on missions, but also having their own story lines that dovetail with that of the main avatar.

At about level 10 of the 55 levels of advancement in SWTOR, each avatar gains a spaceship, a different model for each of the eight classes of characters, useful for traveling between the 18 planets and various space stations and special ships where occasional missions take place. These ships are experienced entirely in the PoB manner, and one's avatar walks around the ship, always experiencing terrestrial gravity. Interplanetary travel requires a navigation interface to tell it where to go, but does not directly pilot it. SWTOR did include a space combat component, but at least in its original version was extremely minor. While accessed in the command room of the spaceship, it was completely unrelated to the structure of the ordinary mission stories. It was like a traditional space combat arcade game, flying a fighter craft along a set path—called *on rails* among gamers—and shooting quickly at many small targets. At the very end of 2013, SWTOR added an expansion called Galactic Starfighter that allowed players to pilot more capable combat craft similar to jet fighters.

As with the PoB ships of SWG, the ships in SWTOR are depicted like the interiors of sea-going vessels, even including stairways, an occasional conventional-looking bedroom, or a meeting room for discussion among crew members that features in a few cutscenes. Each of the companions can be found loitering at a set location, and that is where many cutscenes of their individual stories take place. Each ship has an additional rather limited companion, an unctuous droid who does the cooking and functions as a domestic servant. There are no options to place furniture, trophies, or other individually-selected virtual objects in the ships, but there is a storage space; however that same storage space can also be accessed through terminals at main bases on the various planets.

SWTOR depicts interplanetary and interstellar travel as a jump, taking only a few seconds, and with only one exception the 18 planets are in different solar systems. To add some complexity and travel time, a trip to a planet often requires stopping at an orbital space station above that planet, and walking across a lobby to a shuttle that takes one down to the surface. In returning, one may either take the shuttle, or paradoxically hop straight to one's own ship.

The navigation interface offers information about how SWTOR conceptualizes the galaxy. Indeed, upon opening the interface window, one sees a picture of an entire spiral galaxy, with different sectors outlined that one may select. Clicking a sector opens a simple map of a small number of available destinations, some of which are space battle areas, but most prominently the planets. Figure 9.1 lists the planets, along with one-world descriptions taken from STWOR's atlas, and information from the navigation system acquired when one of my ships was at the main space station for its particular faction. As in the two fantasy MMOs and SWG, planets are depicted as types of terrestrial environments, through which an avatar can run on missions without wearing any special equipment to cope with a hostile atmosphere. The galaxy is divided into spheres of political influence, especially the rival factions of the Republic associated with Jedi knights and located in the Coreward worlds, the Empire associated with Sith like Jedi but on the Dark Side of the Force, and the lesser cartels of slug-like Hutts.

There is a famous scene in the original *Star Wars* movie, when Han Solo uses the word *parsec* to describe speed rather than distance, and Obi-Wan Kenobi's face seems to express slight dismay. Of course *parsec* is a technical term in astronomy, representing about 3.26 light years of distance. It is the distance from our solar system at which a star in the plane of the Earth's orbit will exhibit a parallax shift of one second of arc, as our planet moves from one extreme point to another over the period of half a year. This was a good choice for a measurement, because it could be stated precisely without knowing the exact diameter of the Earth's orbit, but it only works for nearby stars that exhibit significant parallax, and thus it has a distinctively quaint quality.

The parsec distances in Table 9.1 are quite plausible distances, if indeed an STWOR parsec is the same as our parsec, because they cover roughly the width of a real galaxy just as depicted in a ship's navigation interface. The cells with question marks identify six trips that cannot be taken, because a member of one faction cannot visit the low-level planets of the opposite faction. The great distances between the 18 planets support the plausibility that their environments are very Earthlike, because crossing those distances one would pass many other planets that were not Earthlike. One discrepancy is that Nar Shaddaa is not a planet, but a moon of the planet Hutt, yet the distance between them according to the ship navigation system is 678 parsecs, vastly more than the width of any real solar system, yet slightly less than the 687 parsec difference in their distances from the Imperial fleet.

The two fleets are essentially identical in design, with only minor decorative differences. The heart of each is a huge, circular space station. Although wheel-shaped, these stations do not create artificial gravity by rotating, and avatars walk around them and the nearby ships as if normal Earth gravity were in effect. They

Table 9.1 The galaxy in *Star Wars*: *The old republic*

Planet	Terrain	Sector	Distance from republic fleet	Distance from empire fleet
Coruscant	City	Coreward worlds	2,069 parsecs	?
Tython	Mountains	Coreward worlds	3,180 parsecs	?
Ord Mantell	Plains	Coreward worlds	5,358 parsecs	?
Alderaan	Mountains	Coreward worlds	5,607 parsecs	17,076 parsecs
Balmorra	Plains	Coreward worlds	6,651 parsecs	17,236 parsecs
Corellia	City	Coreward worlds	7,454 parsecs	20,607 parsecs
Ilum	Ice	Unknown regions	9,610 parsecs	26,071 parsecs
Taris	Jungle	Seat of the empire	12,203 parsecs	10,207 parsecs
Makeb	Mesas	Hutt space	14,928 parsecs	18,973 parsecs
Quesh	Swamp	Hutt space	16,833 parsecs	11,457 parsecs
Nar Shaddaa	City	Hutt space	17,968 parsecs	15,626 parsecs
Hutta	Swamp	Hutt space	?	16,313 parsecs
Belsavis	Ice	Distant outer rim	18,624 parsecs	33,903 parsecs
Hoth	Ice	Distant outer rim	19,163 parsecs	36,212 parsecs
Voss	Mountains	Hutt space	20,650 parsecs	2,499 parsecs
Korriban	Desert	Seat of the empire	?	2,298 parsecs
Dromund Kaas	Jungle	Seat of the empire	?	1,552 parsecs
Tatooine	Desert	Distant outer rim	22,324 parsecs	28,805 parsecs

are hubs of activity, often each having more than a hundred players on board, socializing in a multi-level cantina or doing business with non-player characters who are skill trainers and vendors. A player can access the ship's storage space directly from several locations on the station, and one area has storage access for the guild to which the player belongs, a group of perhaps a few dozen players who often team up for missions, have their own private text chat channel, and share some resources. Especially important in the economic life of SWTOR is the access area on the station to the game's market, where players offer virtual goods for sale, including raw materials for manufacturing items like weapons, armor, and consumable resources like health potions.

9.4 Science Fiction Universes

Two tremendously interesting but less popular MMOs depict living in extraterrestrial environments more realistically, *EVE Online* and *Entropia Universe.* I included a chapter about each in my 2011 book, *The Virtual Future*, showing how computer-based online virtual worlds might become the venue for future-oriented experiences, substituting for the actual vacuum of outer space (Bainbridge 2011). Both are highly international, *EVE* having Iceland as its home, and *Entropia* having Sweden. They are especially useful here, because they contain the functional equivalent of questionnaires. Both MMOs encourage players to join long-lasting groups, what many MMOs call *guilds* but are *corporations* in *EVE* and *societies* in *Entropia.* Both games give such player groups a system for advertising themselves to new players, and proclaiming their values to all players. A component of that system in both MMOs is a pre-determined list of goals, from which the group leader can select, and which will help new players select which group to join.

EVE Online depicts a galaxy, called New Eden, that was reached hundreds of years before through a wormhole called EVE. Expeditions from Earth began to colonize planets, when the wormhole collapsed, triggering collapse of the interstellar civilizations in New Eden. Now, four competing societies have arisen, and the player initially belongs to one of them, each having its own history, culture, and spaceship designs. The action does not take place on planets, but in outer space, and a major activity is harvesting mineral resources from asteroids, with which to build spaceships and weapons. Experienced players form corporations, begin to colonize additional solar systems outside the regions dominated by the main societies, and enter into very complex conflict with each other. Thus, joining the right corporation is a crucial decision for any player who aspires to ascend beyond the original starting levels.

A website called evewho.com tabulates information about players' characters and corporations, including a listing of the ten corporations with the largest membership at any given moment. On October 16, 2013, Brave Newbies was the largest, with 4,558 members, a group founded at the end of January 2013 for new players, and possibly less significant than the number of its members suggests.

Also intended for new players, but more significant, is EVE University with 2,181 members, which describes itself thus on the website: "For 9 years, EVE University has trained players, and taught them what they need to know in order to be successful in New Eden. We always have positions available for both students and experienced teachers who are looking to pass their experience on to new generations." The self-descriptions of two of the other top ten, Red Federation with 3,471 members and Blue Republic with 3,177, indicated they belong to a player-created political contest:

> Red Federation: The exiles. Driven from their home world for refusing to accept the values and beliefs that were once forced upon them. The members of the newly formed Red Federation stand now against those that they once called brothers and sisters, prepared to defend their newfound freedom. Never again will they live under the tyrannical rule of the Blue Republic. Never again shall a Red bow his head to their ancient religions and hollow prophecies. This day is our day. A day to forge a new way ahead. And any who would stand in our way shall feel our wrath. Ad astra per aspera!
>
> Blue Republic: The Reds claim to be the seekers of truth and enlightenment yet engage in nothing but debauchery. They commit many crimes and must suffer for their deeds, but nothing they have done or indeed can do can be as horrifying as their desecration of the Teapot of Justice, using it to make their so called "Pot Noodle". Our power lies in the shadows, in manipulation and control, in strict obedience to the ancient scrolls, they swore the oath that they now defy, they must die.

GoonWaffe, with 3,350 members, is part of the larger Goonswarm Federation, widely regarded as the most powerful player-versus-player coalition, seeking domination over the others and requiring new recruits to go through an elaborate vetting process, in part to discourage counter-agents from joining under false pretenses. Wildly Inappropriate, with 908 members, is also a member of Goonswarm. Dreddit, with 3,076 members, "is a group of Eve players from the news aggregation website reddit.com". Running with Knives, having 1,094 members, says practically nothing about itself online and discourages strangers from joining. Kung Fu Roguer, with 549 members, and Light of the Moon, with 531, are both Chinese groups.

Table 9.2 shows the results of searching the corporation advertisement system over a period of 5 months beginning in February, 2013, tabulating which of the pre-defined descriptors 553 corporations selected, without their being limited to any particular limit of descriptors. The largest had 575 members, 25 had at least 100 members, whereas 54 had only one member. Two of the large corporations just described also turned up in the search, but are not included in the table. EVE University with 2,272 members listed only New-Player Friendly, as its sole descriptor. Red Federation with 3,284 members when it was checked in June listed Bounty Hunting, Piracy, and Small-Scale Groups, as well as both New-Player Friendly and Roleplay. Blue Republic did not turn up in the search, and indeed an unknown but probably large fraction of corporations do not advertise but recruit members through direct interactions. Indeed, the corporations that were found represent only something like 5 % of the total EVE population, so the results provide only a very rough notion of the distribution of activities and motivations. The two columns of percentages in Table 9.2 give the factions of corporations listing a

Table 9.2 Goals advertised by 553 player groups in *EVE Online*

Goal	Description	553 Corporations (%)	16,925 Members (%)
Player-versus-environment (PvE)			
Exploration	Finding concealed sites or objects in space	40.0	39.2
Mission running	Performing pre-scripted story-based activities, assigned by non-player characters belonging to the four societies or other fictional organizations	45.9	37.9
Incursion	Skirmishes against invasions by non-player spaceships	26.9	22.6
Player-versus-player (PvP)			
Alliance warfare	Conflict between groups of corporations allied with each other	25.1	35.6
Bounty hunting	Attacking ships belonging to wanted criminals	11.6	8.8
Factional warfare	Open conflict among four interstellar societies (Amarr, Minmatar, Caldari, Galente), with the player's corporation belonging to one	16.8	16.3
Piracy	Attacking other players for the loot gained if their ship is destroyed, or to force them to pay ransoms	25.9	27.2
Small-scale gangs	Combat between small groups of players, often lacking much experience, that can prepare them for large-scale warfare between large corporations later on	48.6	58.1
Trade and industry			
Trade	Buying and selling loot and manufactured goods, through a market distributed across the civilized solar systems	17.5	7.9
Mining	Extracting, mainly from asteroids, the resources required to manufacture spaceships and weapons	29.8	18.3
Research	Improving "blueprints" used to manufacture machinery, not to improve the product's performance but to reduce the cost in raw materials required to produce it	14.6	9.9
Manufacturing	Assembling components and raw materials to build a wide variety of products, requiring expensive blueprints and training that takes many days but can be performed by an avatar while the player is offline	18.6	13.3
Other			
New-pilot friendly	The corporations that advertise especially aim their messages at new players, who generally finish many pre-determined tutorials before joining an advanced corporation	53.7	49.8
Roleplay	MMOs are generally role-playing games, but this aspect is less significant in EVE after the early pre-scripted missions	13.4	9.4

descriptor, such as the 221 or 40.0 % of the corporations who listed Exploration, plus the fraction of the total membership represented by these 221.

The three main categories are the standard classification of activities in MMOs. *Player-versus-environment* (PvE) involves exploration and combat missions against non-player characters in the stories. *Player-versus-player* (PvP) is direct combat between players, while Trade and Industry are non-violent activities, which may be between players, but also can simply involve building weapons and spaceships for later use in either PvE or PvP activities.

EVE Online is often described as a PvP sandbox game, because few pre-scripted missions exist beyond the early tutorial levels of experience, and the predominant activities for advanced players concern competition against other players, often involving violent battles. On July 28, 2013, the most massive battle to that date took place between two of the largest alliances of corporations, Test Alliance Please Ignore (TEST) which included the Dreddit corporation, and the ClusterFuck Coalition (CFC), totaling 4,070 spaceships and players. Led by Goonswarm, the CFC had invaded TEST's territory, and triumphed in combat reported to have lasted 5 h and destroyed 2,900 ships. The frankly somewhat radical names of the winners suggest how "masculine" and emotionally intense the battles become in *EVE*. Destruction of a ship does not kill the player piloting it, however, because each "capsuleer" will survive in an escape capsule, losing all the investment in the costly ship unless it was fully insured, but able to fight again another day (Phillips 2013). Early in 2014 all records were broken by a 21-h mass battle involving 7,548 players with destruction of virtual ships estimated to be worth $310,000 (Drain 2014).

Table 9.2 indicates that corporations specialize in a wide range of activities beyond PvP, and two fifths of them advertise Exploration as a main activity. This would seem to suggest a harmony between *EVE* and real space exploration programs, although the corporations that advertise are probably less involved in the PvP than the ones that don't advertise and that are worried about being infiltrated. Yet, if we think through the real history of spaceflight, we realize that it, too, was a case of PvP, reflecting the Cold War between the United States and the Soviet Union. Astronauts were military personnel, and the fact they did not kill each other merely reflects the cold part of the term *Cold War*. Since *EVE's* battles do not really kill people, both fictional and real space programs were symbolic bloodbaths conducted in an outer space that represented a terrestrial battlefield.

Entropia Universe is different from *EVE Online* in at least two respects. First, almost all the action takes place on the surface of one planet, named Calypso, rather than in interplanetary space as in *EVE*. Second, the emphasis is on economic competition, rather than military, in the context of colonization presented as somewhat real. *EVE* is a subscription game, requiring set, periodic payments by players, while *Entropia* is free-to-play, pay-to-win in which one can at any time convert dollars or euros into Project Entropia Dollars (PEDs). During the time I explored Calypso, the exchange rate was 10 PED for one US dollar. As in the case of Linden dollars in the non-game virtual world, *Second Life*, *Entropia's* currency is convertible both from and to external, real-world currencies, so in principle one can profit from setting up businesses on Calypso. For example, major towns have shopping malls, where players can rent space for stores in which they can sell items they crafted, and there are machines in the banks that access a general market. Famously, the gamer news media have reported several rather costly investments, the most recent of which involved the purchase of an entire moon:

> How would you like to own the moon? In the real world, it's not really possible. In *Entropia Universe* it's quite possible and has in fact already happened. A group of players purchased a moon for $150,000 as part of the game's 10-year anniversary celebration. But it's not just a super-expensive piece of property, it's a chance for the owners to make something unique and long-term in the game world. MindArk, the company behind *Entropia Universe*, offered the moon as a chance for players to shape something freely while sharing profits earned from the new location. That means that the investors have essentially licensed out a part of the game for their own development and customization even as it remains a part of the overall gameworld (Lefebvre 2013).

In online gamer forums, critics of *Entropia* often describe it as an online casino, and sometimes express concerns that people susceptible to gambling addictions may lose huge sums in it. Yet frankly I find it one of the very most interesting online virtual worlds, and explored all corners of its virtual territory in the period 2007–2009, at the cost of only about $100. Designed for adults rather than for children, a degree of discipline is required to gain the most from the least investment. As in *EVE*, cooperation is an important element, and players create long-lasting groups, called *societies*.

Because no subscription is required, each avatar is effectively immortal, and societies last forever unless the membership explicitly defects and closes it down. In December 2010, I used the interface in which societies advertise themselves to examine the mottos that expressed their distinctive values, and the many Earth languages they spoke. Fully 3 years later, I logged back in and found my avatar in the exact location I had left him, ready to use the same tools to study a different set of data about the societies. I was able to do a survey of all 787 societies that had at least 10 members, for a total membership of 26,946. There is a 200-member limit, and one part of the interface lists the 50 largest societies, the smallest of which had 177 members.

In the form displayed by the interface, each society must select one of the 14 "hierarchy templates" listed in Table 9.3. These are purely symbolic, because the power structure of all societies is arranged in a set of six classes, each higher class having more powers than the ones below it. For example, the Military hierarchy uses these five names for the ranks, in descending order of power: General, Colonel, Captain, Sergeant, Private, and Recruit. Typically, the lowest rank has no power at all, and is a provisional status given new members until they prove themselves reliable, in which case they are promoted to the second rank, or if they prove unreliable, they may be ejected from the group. The Corporate ranks are also familiar terrestrial statuses: CEO, Vice President, Division Manager, Department Manager, Employee, and Trainee. The term "Order" refers to mystical or religious orders, and its five ranks are: Master, First Circle, Second Circle, Third Circle, Fourth Circle, and Initiate. The fact that Military, Corporate, and

Table 9.3 Distribution of the member hierarchy systems in *Entropia Universe*

System of ranks	787 Societies (%)	26,946 Members (%)
Military	34.6	38.0
Corporate	12.5	9.9
Order	10.9	11.0
Hunters	7.4	6.9
Mercenaries	7.2	6.7
Basic	6.4	7.3
Explorers	5.7	5.0
Outlaws	4.3	3.7
Security	3.0	4.5
Miners	2.3	2.4
Mystics	2.3	1.2
Scientists	1.9	2.0
Traders	1.3	1.1
Crafters	0.3	0.2

Order are the three most often used implies that Entropians apply very traditional distinctions from terrestrial society, namely the military, capitalist, and religious systems of power and status.

The less popular status systems are mostly self-explanatory, although Basic is simply the default with essentially meaningless terms for the statuses, Security is the police or city guards, and the very unpopular Crafters are skilled artisans. Perhaps most notable is the low popularity of Scientists among the status systems, the ranks of which are: Professor, Assistant Professor, Senior Researcher, Researcher, Junior Researcher and Student.

While the hierarchy templates are purely symbolic, the society format includes two more substantive questions, concerning which two activities the society considers its primary and secondary activities. As with the hierarchy templates, these two choices are selected from a list presented by the user interface in a somewhat unusual way for a questionnaire. The standard method for an item in an online questionnaire for which the user is limited to one answer from a list is *radio buttons*. This term is by analogy with old-fashioned car radios, in which there is a row of pushbuttons, each set to a different radio station frequency, and pressing one unselects the previous selection. But radio buttons take up a lot of space on a computer-administered questionnaire, so Entropia uses a small window in which there initially is the selection "None," framed by two clickable arrows. Selecting the right-hand arrow once replaces "None" by "Mining—All." Clicking again goes to "Mining—Energy Matter," and once again, to "Mining—Minerals," the two specific kinds of resources that can be mined on the planet Calypso. Of the 787 societies, 97 selected "Mining—All" as one of its two activities, 2 selected "Mining—Energy Matter" which is the extraterrestrial equivalent of fuel oil, and 11 selected "Mining—Minerals." In Table 9.4, I combined these mining categories, to give a total of 110 societies, or 14.0 %. In a few cases, a society picked the

Table 9.4 Distribution of society activities in *Entropia Universe*

Activities	787 Societies (%)	26,946 Members (%)
Functional		
Hunting	38.6	33.2
Mining	14.0	13.7
Manufacturing	4.1	2.0
Aspirational		
Having fun	33.7	30.0
Personal development	18.4	22.1
Be helpful to newcomers	17.8	23.9
Exploring the new world	12.2	12.4
None	9.1	1.6
Survival	7.6	5.6
Partying	5.3	4.4
Commercial business	4.6	4.6
Fight the power	4.4	3.0
Fight evil	4.1	4.6
Trading	3.9	3.1
Keeping the peace	3.4	3.8
Protect the meek	2.5	2.9
Fame seeking	2.5	3.4
Industrial business	2.3	3.0
Research	2.2	2.5
Conserve nature	2.0	2.6
Offer security	1.9	1.7
Landowning	1.8	1.6
Uphold the law	1.5	1.3
Join the highest bidder	1.3	0.9
Mindforce usage	0.6	0.4

same activity twice, so it was counted twice. Clicking the left-hand arrow from "None" would select "Join the highest bidder," which implied people would join whichever society rewarded them most for doing so.

Like Mining, Hunting and Manufacturing offered subcategories, and these are the main activities that earn a player resources from the environment, so Table 9.4 tabulates them separately as *functional* activities, which all players must do in order to prosper on Calypso. The others I call *aspirational* activities because they are optional and reflect different motivations that players may have. "Having fun" may seem trivial, and in Chap. 4 we saw it is a real but unconvincing justification for spaceflight. "Exploring the new world" is a significant category, but less popular than the economic activities of Hunting and Mining. "Personal development" refers to advancement upon Entropia's extremely complex system of experience levels, represented as a huge number of technical fields in which one may gain expertise.

As *EVE* is a PvP sandbox, *Entropia* is a PvE sandbox, emphasizing cooperation between players to deal with the hazards and opportunities in the extraterrestrial environment. Violent combat between players is possible but unimportant. Economic competition is far more important, but an economic market is a system of cooperation. The planet Calypso is very earthlike, which is consistent with the indication that it is several light years away, and the next step in colonization after the solar system. Remarkably, the online history of this fictional colonization begins with a rather compelling analysis of how real colonization might come about:

> Curiosity and greed opened space to mankind. As he looked up in the night sky, curiosity sparked Man's ability to dream of space. Greed gave him a reason to invest in the technology to reach the stars. The humble single-use rocket would probably never have been used for space travel unless nations used them as weapons of war. Reusable rockets and space shuttles were deployed when war no longer loomed over the skies of Earth and civilian governments demanded more for less. The space plane made space travel feasible for corporations, and space as a market for all commercial interests opened up.
>
> "The giant leap for mankind" was dwarfed by another step: the permanent habitation of space. The journey to Earth's lunar companion was a short excursion; living in space is another matter, and living in space independently of Earth is on an order of magnitude more important. With access to space and the promise of ever-expanding wealth and profit, there was a great need for the ability to live in space without having to depend on expensive resources lifted from Earth—to truly live off the land.
>
> From the ability of cheaply climbing the gravity wells and the ability to live off the land, Mankind was able to secure her destiny as a space-faring species. Soon, the planets and moons in the Solar system were colonized and the era of interplanetary conquest had begun (www.planetcalypso.com/planet-calypso/the-story-of-calypso/).

We may, of course, doubt whether cheap access to outer space will ever be possible, and whether exploitation of extraterrestrial resources would ever be profitable. But given those two assumptions, space could become the venue for most of humanity's traditional motivations and activities. *EVE Online* emphasizes military-style competition, and *Entropia Universe* emphasizes the economic alternative, but both require human cooperation and exploitation of natural resources. In these features, both games are rather realistic, representing rather logical forms of science fiction that allow people today to experience something like the interstellar societies of tomorrow.

9.4.1 The Harsh Truth

The simulated planetarium, *Starry Night*, is a high quality educational tool for astronomy students and guiding ephemeris for amateur astronomers. Yet software that depicts the planets realistically is not nearly so widely popular as videogames and online virtual worlds that depict extraterrestrial locations as Earthlike environments suitable for ordinary human activities. One could argue on the one hand that the six MMOs described here encourage public enthusiasm about the real space program, but at the cost of grossly misrepresenting it. Or, one could argue that the popular requirement for traditionally human experiences in space erodes real exploration, because it does not match the conditions of the actual galaxy around us.

A very different perspective would suggest that the marriage of spaceflight and information technology constitutes the real future of spaceflight. Humans will remain on Earth, enjoying popular science fiction in their mundane lives, and gaining some appreciation for real discoveries about the universe from it. We will not need a radically new launch technology, because automated probes can already be sent everywhere in the solar system, and the needed advances will be in areas such as robotics and sensor technology, so the probes can carry out ever more sophisticated scientific research.

Still, no dedicated science fiction fans, regardless of their favored media, could really be satisfied by the harsh judgment that the heavens are closed to human habitation, and only our thoughts can soar beyond the bonds of Earth's gravity. We may then wonder what kind of social, economic, and cultural transformation might put our species in a position to escape this terrestrial prison.

References

Anonymous. (2009). *Starry night: User's guide* (p. 7). Toronto, Canada: Simulation Curriculum Corporation.

Bainbridge, W. S. (2010). *The warcraft civilization*. Cambridge, Massachusetts: MIT Press.

Bainbridge, W. S. (2011). *The virtual future*. London: Springer.

Bronner, F. (1999). *Buzz Aldrin's race into space: Rules of play*. Irvine, California: Interplay Productions.

Burroughs, E. R. (1917). *A princess of Mars*. Chicago: A. C. McClurg.

Burroughs, E. R. (1946). *Escape on Venus*. Tarzana, California: Edgar Rice Burroughs Inc.

Drain, B. (2014). *EVE evolved: The bloodbath of B-R5RB, Massively*, February 2. Retrieved February 2, 2014, from massively.joystiq.com/2014/02/02/eve-evolved-the-bloodbath-of-b-r5rb/.

Farkas, B. (1998). *StarCraft: Prima's official strategy guide*. Roseville, California: Prima.

Garfield, O. (1955). *Geniacs: Simple electronic brain machines, and how to make them*. Retrieved January 25, 2014, from www.computercollector.com/cgi-bin/exec/compcol/menu-st.cgi?directory=/archive/geniac/manual.

Gygax, G. (1979). *Advanced Dungeons and Dragons, Dungeon masters guide*. New York: TSR/Random House.

Heinlein, R. A. (1949). *Red planet, a colonial boy on Mars*. New York: Scribner's.

Lefebvre, E. (2013). *Entropia Universe players buy the moon, Massively*, April 4. Retrieved February 17, 2014, from massively.joystiq.com/2013/04/04/entropia-universe-players-buy-the-moon/.

Lummis, M., & Kern, E. (2007). *World of warcraft: The burning crusade*. New York: BradyGames.

Marks, R. B. (2003). *EverQuest companion* (p. 53). Emeryville, California: McGraw-Hill/Osborne.

McCubbin, C. (2005). *Star Wars Galaxies: The complete guide*. Roseville, California: Prima.

McCubbin, C., Ladyman, D., & Frase, T. (Eds.). (2005). *Star Wars Galaxies: The total experience*. Roseville, California: Prima.

Phillips, T. (2013). *This is what the largest online space battle in history looks like, Eurogamer*, July 29. Retrieved December 30, 2013, from http://www.eurogamer.net/articles/2013-07-29-this-is-what-the-largest-online-space-battle-in-history-looks-like.

Searle, M. (2011). *Star Wars: The Old Republic explorer's guide*. Roseville, California: Prima.

Swenson, L. S., Grimwood, J. M., & Alexander, C. C. (1966). *This new ocean: A history of Project Mercury*. Washington, D.C.: National Aeronautics and Space Administration.

The history of Planet Calypso. Retrieved December 30, 2013.

Chapter 10
Frontiers

Opinion leaders and the general public hold a wide variety of views about the meaning of spaceflight, but we have seen no evidence that citizens of the United States demand major advances at the present time. Aside from causing sadness for those of us committed to human exploration across the wide universe, this realization may unleash justified fear that all kinds of progress may be coming to an end. With special emphasis on the frontier theory of Frederick Jackson Turner about what happens when a frontier closes, this chapter will not merely summarize the insights that can be gleaned from earlier chapters, but ascends to the more general question of what the meaning of spaceflight for the general public says about the human future.

In the first chapter we noted the technological determinist view that a renaissance of space exploration could come about after other fields of science and technology had independently established the preconditions for new and more efficient launch systems. At that point, some decades in our future, new motivations might be found to energize a second wave of exploration and perhaps even colonization of the Moon and Mars. Especially if economic and demographic social problems had called into question aspects of traditional culture, significant fractions of the general public might come to believe the best proving ground for general human progress might indeed be far above the surface of our small planet.

Yet it is also possible that we have reached a boundary in evolution, when humanity has expanded geographically to its maximum scope, and any further progress will consist of improvement in living conditions and science-based wisdom, here on this one small planet. Transitioning to a stable world situation may prove exceedingly difficult, as natural limits to growth will result in environmental degradation, increased conflict, and vulnerability to both economic and psychological depression. Perhaps our experience failing to conquer space will provide the wisdom needed to see alternative paths for progress.

W.S. Bainbridge, *The Meaning and Value of Spaceflight*,
Space and Society, DOI 10.1007/978-3-319-07878-6_10

10.1 Closing of the Frontier

Near the middle of the twentieth century, American computing pioneer and presidential advisor Vannevar Bush called science *the endless frontier* (Bush 1980; Zachary 1997). Then president John F. Kennedy included space exploration in his New Frontier. Spaceflight enthusiasts may be wrong to believe the Apollo Program really was an integral part of Kennedy's vision, however. The Wikipedia page for the New Frontier quotes from the presidential nomination acceptance speech Kennedy gave, July 15, 1960, that describes the frontier in terms of "uncharted areas of science and space, unsolved problems of peace and war, unconquered problems of ignorance and prejudice, unanswered questions of poverty and surplus." Then it cites historian Robert D. Marcus: "Kennedy entered office with ambitions to eradicate poverty and to raise America's eyes to the stars through the space program."

After these quotations from political rhetoric, however, the article says nothing more about the space program, instead offering a very long list of legislative accomplishments primarily in social areas: economy, taxation, labor, education, welfare, civil rights, housing, unemployment, health, equal rights for women, environment, agriculture, crime, and finally defense ending with the Vietnam War and the abortive Bay of Pigs invasion of Cuba. Wikipedia is an expression of popular opinion, because the articles are created by anonymous volunteers. As of February 14, 2014, the page had been viewed 7,539 times in the past 30 days, and had received a total of 598 revisions since being created March 29, 2004. None of the editors apparently saw fit to add a section on the Apollo Program, or even to insert a prominent link to Wikipedia pages about it. The page restricted "New Frontier" to Kennedy's "domestic and foreign programs."

Star Trek called space the Final Frontier, perhaps inadvertently suggesting that progress would be impossible without spaceflight. Then, four decades ago, Apollo proved to be a different ancient myth from one about the Greek sun god, more like Icarus who flew too high and fell back to Earth. What, in general, is the meaning of a frontier for society, and can we progress without one?

A century ago, Frederick Jackson Turner argued that the American frontier stimulated democracy and economic progress, and that the closing of the frontier around the 1890s endangered these advances (Turner 1920). Whatever cultural advantages Europe may have possessed, its chief advantage was its geographic location. First, the peoples of the Mediterranean built economies and technologies based on trade within that ocean-sized lake, then the nations on the Atlantic employed the Mediterraneans' advances first for trade along their own seacoasts, then across the great oceans in an Age of Discovery that promoted science and technology as well as exploitation of ever-greater natural resources. Turner noted that residents of the American frontier were free to govern themselves, and this freedom spread back to the lands the colonists had come from. But economic progress also produced freedom and accelerated innovation, because it constantly undercut old authority structures and funded experimentation with new ones.

There are two problems applying this frontier metaphor to spaceflight. First, the Wild West was a new area for colonization, yet even a dedicated spaceflight scholar like Roger Launius doubts that humans will be able to colonize the solar system without transforming our very fragile physical natures (Launius 2010). Second, Turner's thesis required that the people exploring and exploiting the frontier were doing so as free individuals, not as agents of a centrally controlled government. Yet spaceflight is too expensive for individuals and small groups to achieve, and a wagon train to Mars would find far too hostile an environment to set up a town and start planting crops. Furthermore, human history provides an earlier example of a frontier that was explored not by free individuals but by governmentally controlled forces, and that had a sad conclusion. I refer to the expansion of the Roman Empire.

Russians and Germans and Scots may well feel they live in bountiful lands with long histories of civilization, but for the ancient Romans, they and their territories were contemptible wastelands and not worthy of colonization. So here is a different metaphor. Our entire world today can be represented by ancient Greco-Roman civilization, that ceased expanding once it had conquered the entire circuit of the Mediterranean. We seem to have ceased expanding, now that we have occupied the entire circuit of Planet Earth. This part of the metaphor is optimistic, because we know that the Americas and all the rest of our globe now belong to world civilization, even though the Romans never crossed the Atlantic, so the solar system may be conquered in some future Age of Discovery, if not very soon. The Americas may become Mars. With a touch of humor, we can even specify what was the equivalent of the Moon, for the Roman expansion program, a place that was reached and then abandoned: Britain.

Turner would say that the Roman example confirms his frontier theory, because Rome quickly descended into dictatorship and economic decline, once its boundaries solidified (Gibbon 1880). I would debate with him about this on two related grounds. First, Rome was still expanding when Caesar and his heirs established the imperial form of government. Second, the expansion was led by massed armies not by free pioneers, and thus the expansion was an effect of tyranny, rather than its conclusion being the cause of tyranny. Had Rome grown more naturally, then its expansion would have been led by farmers and traders, with new lands being added to the government only after they had been settled, as was the case for states in America. But the really pessimistic aspect of the Roman case is precisely that it was followed by a 1,000-year Dark Age.

This raises the possibility of the fall of our own civilization. Like technological determinism, this theoretical perspective dates back more than a century, and has faded from the social scientific journals. The one variant that has been discussed widely in recent years is what Samuel P. Huntington called the *clash of civilizations*, the thesis that the world could not unify because humanity possessed several very different cultures, each with its own distinctive values, of which technocratic capitalism was only one (Huntington 1996). As demographer Nathan Keyfitz observed, Western Europe and the United States enjoy democratic freedoms and achieve technological innovations, while Islam has less of these qualities but possesses a family structure that sustains a much higher birth rate (Nathan Keyfitz 1987). Thus these two cultures will not only exist significantly apart from each other, but also inevitably clash, according to this theory. Sociologists of religion have observed that faiths that compete with each other tend to accentuate their differences, so it is conceivable that during the twenty-first century some of the clashing civilizations will support space exploration, for example American and Chinese (Niebuhr 1929). But that assumes that the technologically advanced societies will survive.

Theories of the rise and fall of civilizations were popular over the decades that technologically determinist scholars like Childe, Gilfillan and Ogburn wrote from a more optimistic perspective. I believe those theories are given little consideration today for two reasons, neither of which should be decisive. First of all, it is very difficult to evaluate a theory about the rise and fall of civilizations, because there are so few cases to analyze, and we lack the material and ethical basis for carrying out rigorous experiments on them. Second, the civilization theories tended to be perceived as politically very conservative, thus unattractive to social scientists who tended to stand near the left end of the political spectrum, and technologically pessimistic, thus not attractive to leaders of industry either. Those problems should not deter us here, because we seek very general theories, as well as more narrow and thus more testable hypotheses, and because spaceflight by its very nature inspires us to transcend terrestrial political battles even as we leave the Earth itself.

One of the most influential theorists prior to the Second World War, but whose works were discredited by the fact that some in the Nazi leadership admired them, was Oswald Spengler (Spengler 1926). In *The Decline of the West*, he argued that each major civilization was based on a fundamental idea, and *space* was the concept that had given meaning to Western Civilization. On one level, *space* in Spengler's writings refers to the urge for geographic expansion that led Britain, France and Spain to establish colonies all around the world. But at times he also uses astronomical metaphors, and at other times says that the West has a general compulsion to become boundless. However, discouraged as many others of his generation were by the First World War, Spengler exudes pessimism, believing that the West is losing its ambitions, and without them it will collapse.

Pitirim Sorokin, who argued against Spengler even as he shared some of the same assumptions, had a different experience of the First World War, being a Russian rather than a German and even playing a supportive role in the interim government that took power temporarily after the Czars only to be swept aside by Marxist-Leninism (Sorokin 1937). He barely survived the Revolution, then came to the United States where he founded the Department of Sociology at Harvard University. Sorokin agreed with Spengler that every civilization is based on specific ideals, but they may not be simple, and a civilization could rise again after a fall. At its birth, Sorokin asserted, each civilization is *ideational*, based on transcendent principles that are often religious in form. The rise of a civilization can be conceptualized, therefore, as the success of a militant social movement growing out of but somewhat separate from the local traditional culture. It is led by fanatics whose use their ideology to recruit and control followers, but the fanaticism will fade over the decades, as the culture becomes *sensate*, which was Sorokin's way of saying materialistic or hedonistic. As the faith that gave original meaning to the civilization erodes, so does the society's unity and sense of purpose, leading eventually to collapse.

The implications of Sorokin's theory for technological progress are clear but slightly complex. As a culture becomes more sensate, it may also become more technological, so a culture that emphasized technology among its founding principles will appear to be progressing ever more rapidly, when it is really accelerating toward a precipice, from which the fall may be especially abrupt. Unless some other civilization sweeps into confiscate the wreckage, however, the Dark Ages that follow the fall of a civilization can become the breeding ground for new social movements, one of which will establish a new civilization on the ashes of the old.

Writing from a British perspective, also disillusioned by the First World War but more optimistic, as befits a citizen of a victor nation, Arnold Toynbee theorized that a civilization was created by an elite class who responded successfully to challenges that confront the society (Toynbee 1947). One of his key examples was ancient Egypt, in which he believed climate change imperiled the early way of life, and the elite that created a nation on the banks of the Nile had to develop a huge range of new technologies to thrive in that environment. Note that both Sorokin and Toynbee emphasize the role of elites, and thus their theories are neither democratic nor technologically determinist. Toynbee's theory that human progress could be understood in terms of *challenge and response*, explicitly stating that only a tiny minority in the society had the ability or position to find a successful response. In the context of our research here, Toynbee would be skeptical we could learn much by polling the general public, would focus on the potential of a renewed spaceflight social movement, and he would give more credence to those of our datasets that examined special minorities such as Harvard students and science fiction fans.

10.2 A Clash of Cultures

Setting aside Huntington's classification of civilizations, we can consider how the spaceflight social movement competes with other cultural traditions within western societies. Most obvious is the distinction between science and the humanities. In the preface to his 1894 spaceflight novel, *A Journey in Other Worlds,* John Jacob Astor IV asserted one of the main justifications for space exploration, in contrast to what the humanities can offer:

> The protracted struggle between science and the classics appears to be drawing to a close, with victory about to perch on the banner of science, as a perusal of almost any university or college catalogue shows. While a limited knowledge of both Greek and Latin is important for the correct use of our own language, the amount till recently required, in my judgment, has been absurdly out of proportion to the intrinsic value of these branches, or perhaps more correctly roots, of study. The classics have been thoroughly and painfully threshed out, and it seems impossible that anything new can be unearthed. We may equal the performances

> of the past, but there is no opportunity to surpass them or produce anything original. Even the much-vaunted "mental training" argument is beginning to pall; for would not anything equally difficult give as good developing results, while by learning a live matter we kill two birds with one stone? There can be no question that there are many forces and influences in Nature whose existence we as yet little more than suspect. How much more interesting it would be if, instead of reiterating our past achievements, the magazines and literature of the period should devote their consideration to what we do NOT know! It is only through investigation and research that inventions come; we may not find what we are in search of, but may discover something of perhaps greater moment. It is probable that the principal glories of the future will be found in as yet but little trodden paths, and as Prof. Cortlandt justly says at the close of his history, "Next to religion, we have most to hope from science." (Astor 1894)

The Professor Cortlandt he cites is an imaginary character in the novel, writing in the year 2000, more than a century after Astor, and in the real 2000 there were reasons to doubt whether religion really could compete with science in terms of benefits for humanity (Bainbridge 2007a). Astor's own future ended with the adventure of dying when the Titanic ocean liner sank in 1912, a reminder that journeys can be dangerous. The most influential and widely debated expression of conflict between science and the humanities was C. P. Snow's Two Cultures thesis dating from 1959 (Snow 1959). Among the most influential recent attempts to counteract cultural divisions was *Consilience*, published in 1998 by evolutionary biologist Edward O. Wilson (Wilson 1998). One could well argue that both the humanities and sciences are segments of the human cultural genome, serving different but equally essential functions.

We have already considered in depth one of the main fields where science and the humanities combine, namely science fiction. For the past 15 years I have been involved in another one, that might be called the Convergence Movement, primarily in my role as a bridge between the social and computing sciences at the National Science Foundation. A first step was participation in a major workshop held at NSF on September 28–29, 2000, just when I happened to be moving from NSF's social science directorate to computer science directorate, on the societal implications of nanotechnology (Roco and Bainbridge 2001). There are two theories of the origins of nanotechnology, first that it was a natural development of materials sciences, and second that it was defined and inspired by science fiction. Unlike many of my colleagues in government and industry, I see considerable merit in the second theory.

Yes, research and engineering at the dimensional scale just above the size of atoms to around 100 nm was bound to happen, but conceptualizing it as a distinct field with enormous potential may have come from science fiction. The term *nanotechnology* was popularized, if not exactly invented, by K. Eric Drexler who had been deeply involved in the spaceflight social movement and may have turned to nano as a way to revive hope after the stalling of the space program (Bainbridge 2007b). His 1992 book *Nanosystems* encouraged many other people to think along these lines, including science fiction writers (Drexler 1992). We do not have good historical documentation at this point about how or even whether scientists and engineers in the National Nanotechnology Initiative were inspired by Drexler and SF, but I theorize that they were influenced yet would be reluctant to say so.

Table 10.1 Connections among science fiction media and spaceflight

Questionnaire item	SF novels	*Star Wars*	*Close Encounters*	*Star Trek*	*Battlestar Galactica*
In the long run, discoveries made in our space program will have a big payoff for the average person	0.25	0.12	0.12	0.19	0.15
The United States is spending too much money on space, so appropriations for the space program should be reduced	−0.24	−0.10	−0.08	−0.16	−0.15
Liking science fiction novels	–	0.31	0.25	0.49	0.33

Once nanotechnology had been conceptualized, and people had begun to consider its societal implications, it quickly became obvious that it connected technically and conceptually with molecular biology and with microelectronics that had begun to build computer chip components smaller than 100 nm. A group centered on Mihail C. Roco at the National Science Foundation then began to work out a wider perspective on the convergence of nanotechnology, biotechnology, information technology and new technologies based on cognitive science (Roco and Bainbridge 2003). As fate dictated, I was centrally involved in all the conferences and book publications of this NBIC convergence—Nano-Bio-Info-Cogno—and then also involved in the 2012–2014 period when the convergence expanded to include the social sciences, and to postulate the future convergence of science with society (Roco et al. 2013).

Launching a social movement does not guarantee it will achieve its objectives, yet creating a new kind of society based on scientific knowledge and engineering idealism may be necessary before humanity is ready for a second wave of space exploration. Science fiction certainly provides much of the cultural preparation, today as it has for decades. Given that we considered SF literature and dramatic media in two different chapters, it is worth verifying that they belong to a connected if diffuse subculture.

In a study I carried out at the University of Washington in 1981, I asked 1,439 students how much they agreed or disagreed with some statements about technology, and how much they liked science fiction novels, two sci-fi movies, and two sci-fi television programs of the period (Bainbridge 1982). Table 10.1 shows the tau-correlations between the five science fiction variables and the two statements, because those two agree-disagree items were ordinal variables, plus the ordinary Pearson's r correlations between the sci-fi items and science fictions novels, because they were on the same interval preference scale. Given the large number of respondents, all these coefficients are statistically significant.

We do see the expected kinds of correlations between science fiction and the space variables, although the literature was a more powerful predictor than the television programs, which in turn were more powerful than the movies. In a different analysis, I compared the percentage who agreed that "appropriations for the space program should be reduced" across the full range of preferences. Those who loved science fiction novels were nearly 40 % points less likely to want funding reduced than those who hated them. The differences across the full range of the preference scales for *Star Trek* and *Battlestar Galactica* were about 20 % points. The validity of these relationships is supported by the very high correlation of 0.49 between science fiction novels and *Star Trek*, given that this classic TV program was historically based on the subculture of science fiction literature.

The previous chapter considered computer-based simulations of spaceflight, notably those that ordinary people can experience in games, although I tend to think that the virtual worlds described in that chapter have evolved beyond the definition of *game*. In reporting research on the highly popular *World of Warcraft*, I suggested that a better description might be *total work of art*, a concept taken from the admittedly radical essay by Richard Wagner, *The Artwork of the Future*, written in 1849 (Wagner 1994). Among the themes of Wagner's career as a composer of grand operas was a reconsideration of the relationship between Paganism and Christianity. Notice that one way of reading Astor's paragraph above is that there are not two but three competing cultures: humanities, science, and religion. During the Middle Ages all high forms of art were based in the church, but the Renaissance liberated art to be secular while remaining sophisticated, which could be construed as the most important way that it resurrected Classical Paganism.

The cultural constellation to which science fiction belongs may be a secular substitute for religion. In *Dimensions of Science Fiction* I wrote: "The sense of wonder afforded by SF is akin to religious awe, and in it people may find at least a taste of the spiritual uplifting and cosmic meaningfulness once gained in church." (Bainbridge 1986a) In 1974, of 1,483 respondents to the General Social Survey, just 6.8 % said they had no religion. In that same year, 130 editors of SF fan magazines (fanzines) responded to a questionnaire I mailed them, and fully 31 % said they had no religion. My 1973 survey of 74 members of the New England Science Fiction Association found that 48 % had no religion (Bainbridge 1986b). This is not to say that science fiction fans have the degree of faith possessed by religious evangelists, but they may have as much hope as can be logically justified in our secular society. I explored this theme in my 2013 book *eGods: Faith Versus Fantasy in Computer Gaming*:

> Religion has always been deeply involved in the creative arts, but the relationships between them are changing. Perhaps we shall come to see religion merely as an especially solemn art form. Suspension of disbelief is the essence of art, according to Samuel Taylor Coleridge, and electronic games are a new and powerful art form that often depicts religion. Yet we may wonder if suspension of disbelief is really very different from belief itself. Traditional religions took their faith very seriously, and pious believers today would be shocked if told their God was not very different from an elf's image on a computer monitor. Yet much may be gained by thinking from that admittedly radical perspective. (Bainbridge 2013)

In other words, humanity may need to preserve the dream of spaceflight, to compensate us for some of the suffering we experience here on Earth, and as a possible long-term goal. It is worth remembering that the Christian Crusades came a thousand years after the crucifixion of Christ, proof that a transcendental social movement may energize human action even after a very long delay. I can almost hear SF fans and rocket scientists wailing in unison, "We cannot wait that long!" Indeed, how could humanity sustain hope for spaceflight more than just a few more years?

10.3 Cultural Stability

In its July 24, 2011, issue, the *Washington Post* newspaper published a cartoon by Tom Toles about the conclusion of the Space Shuttle program. It shows a grounded shuttle, on a cracked Earth with a factory in the background spewing pollution labelled "Earth's climate destruction." A returning astronaut says, "Our final report: There's no escape." A tiny commentator at the bottom adds, "...and no intelligent life down here." This little sketch summarizes a powerful viewpoint shared by many knowledgeable people today: We must utterly transform our civilization to live within the confines of this fragile planet, and spaceflight will have only a very minor role to play.

An entire series of books could be written about public opinion concerning environmental sustainability, economic growth and fairness, and general technological progress. We have already seen some relationships, such as a disconnect between environmentalism and spaceflight, and others might become apparent. Here we can most profitably look back at one of the most powerful explanatory variables, and consider what it might possibly mean: gender.

Men are more enthusiastic about spaceflight than women. Originally, American astronauts were all male current or former military personnel, and the connections to the military-industrial complex favored males. Thus, the masculinity of spaceflight could be an historical accident, the influence of which may fade over time. However, in traditional cultures the two genders played different roles in society, and possessed different but functionally compatible values. Technological determinist William F. Ogburn, who proudly called himself a Feminist, argued that several changes in the practical operation of society were liberating women from their traditional domestic role (Ogburn 1922). Improved health reduced infant mortality, which meant they did not need to have so many children to sustain the human population. Many of the productive tasks performed inside the household during the nineteenth century either became unnecessary or were done in factories, from candle making to the preservation of food. According to Ogburn, cultural lag delayed the liberation of women, but eventually it would be fully accomplished.

We cannot be certain about many issues related to gender differences. If the cultural lag is completely eliminated, will women be entirely liberated from their traditional domestic focus, at the same time men are liberated from the battlefield? Or will there still be differences of degree, with men and women showing some persistent specialization in the percentages in each gender who seek particular activities and hold particular values? We may not have full answers to such questions very soon, and during the current historical situation of uncertainty individuals are free to differ in what they suspect the truth to be. But in many, many polls, the two genders do differ in terms of the percentages who express various opinions, and they also still differ in their daily interests and activities. Thus, we can use gender differences as a proxy for two competing value systems that may be meaningful quite apart from gender, but measured to some extent by current gender differences:

Preservation:Traditionally attributed to women but significant for all human beings, this value system emphasizes health, social stability, and maintenance of a nurturing home environment

Aggression:Traditionally attributed to men but now potentially obsolete, this value system does not merely defend the home but seeks to conquer a wider territory, to increase the power of victors in conflict that can escalate into warfare, perhaps serving to promote human evolution in prior centuries but now threatening human survival.

Neither the reader nor the writer of these words is required to believe them, but they suggest a way of thinking that has implications for spaceflight as well as for transformations of human society that may not entirely be under our control. In a sense, this is an extension of Ogburn's analysis from nearly a century ago. Yes, women are liberated from their traditional domestic role, but men also are liberated from their traditional combat role. The relevance here is that as illustrated by the Space Race, space exploration was a form of aggression. As illustrated by the lack of current motivations for major new efforts, space exploration may not serve preservation, except in very limited ways like monitoring the Earth's environment from orbit, and guarding against wayward asteroids.

Traditionally, science and engineering have been masculine fields, although females have begun to make headway recently. A very comprehensive source of information about American women's involvement in science is a book-sized report, *Women, Minorities, and Persons With Disabilities in Science and Engineering*, issued by the National Science Foundation. The 1999 edition says that in 1966 women received only 8.0 % of all American doctoral degrees in science and engineering, but their proportion had risen to 31.2 % in 1995 (National Science Foundation 1999). Professional women scientists tended to be concentrated in such fields as biology and the social sciences, and they were especially rare in fields more closely related to spaceflight. For example, in 1995 women earned only 11.6 % of American doctorates in engineering, 18.6 % of doctorates in computer science, and 22.9 % in physical sciences-compared with 38.1 % in biological sciences and 63.5 % in psychology. At earlier points in their educations, males and females are more similar. In 1994, for example, 9.1 % of girls graduating high school had taken calculus, only slightly lower than the 9.4 % of boys.

The newest edition of the NSF report, published in 2013, notes: "Overall, more women than men graduate from college with a bachelor's degree; however, men earn a higher proportion of degrees in many science and engineering fields of study (National Center for Science and Engineering Statistics 2013)." Considering

Table 10.2 Education and spaceflight support by gender

Educational attainment	Number of respondents		Too little spent on space	
	Women	Men	Women (%)	Men (%)
Less than high school	586	533	7.0	11.1
High school	2,280	1,838	9.2	19.6
Junior college	393	284	13.7	22.9
Bachelor degree	730	677	14.2	25.7
Graduate degree	366	381	17.2	27.0

the period from 1991 to 2010, the report notes that about two thirds of psychology degrees are earned by women, compared with rather less than a third of degrees in engineering and computer science. The trend in female participation in psychology and social science has been upward, but the overall trend in computer science and engineering had been static, but with a decrease in computer science balanced by a slight increase in other engineer-related fields. Certainly we are very far from reaching gender parity in the fields most obviously relevant to spaceflight.

Some variables that do not correlate with gender add to the effect of gender to determine enthusiasm for spaceflight. An obvious example is social class. For GSS data dating from 2000–2012, just 8.2 % of 329 women who self-identify with the lower class feel space funding is too little, compared with 16.3 % of 141 women in the upper class. For men, the percentages are higher at both social class extremes, 16.5 % of 206 in the lower class and 21.3 % of 136 in the upper class. In this period, the gender difference on spaceflight enthusiasm was rather large. Only 10.8 % of 4,359 women wanted space funding increased, compared with 20.5 % of men. But as we have just seen, the difference between lower-class woman and upper-class men is even greater, 8.2 % versus 21.3 %.

We see a similar pattern for education, which of course is one indicator of social class. Table 10.2 analyzes support for increased space funding for the two genders separated. From women with little education to men with high education, the difference in feeling too little is being invested is great, 7.0 % compared with 27.0 %. But the opposite comparison is also interesting. From men with little education to women with high education, there is a smaller but significant difference, from 11.1 to 17.2 %. Yes the genders differ, but the educational difference seems largely independent. This suggests, through an admittedly uncertain but plausible train of logic, that over time spaceflight could gain new intellectual and preservation motivations, compensating for the loss of aggression motivations.

One theme much discussed within the conferences that led to the 2013 Converging Technologies report was the notion that increased emphasis on science and engineering education could mitigate the current elevated level of unemployment, by giving more people better job skills. Apparently the general public has come to this view over the years. One recent journal article compared responses from the General Social Survey in 1975 and 2002, finding that 8 % of respondents wanted space funding increased in 1975, and 12 % in 2002, which may seem like a small increase of 4 % points, or a big 50 % increase in terms of the baseline of 8. But support for increased funding for the nation's education system was undeniably great, from 51 to 74 % (Plutzer and Berkman 2005).

In the abstract, it is certainly true that improved education can be beneficial, but one can wonder if perhaps the positive effect will be small. Advances in information technology may have two parallel effects that tend to increase unemployment. First, and most obviously, automation eliminates specific jobs, without automatically creating an equal or greater number of new ones. Second, and more subtle, progress in information technology allows a smaller number of very well educated people to accomplish vastly more, thus greatly increasing inequality and rendering unemployable many people with decent but not exceptional levels of skill. This raises a more general question about the meaning and value of education.

Considering the full sweep of formal education in America, Richard Merelman suggested there were three modes of adjustment between public education and social structure, which came into being sequentially but then existed concurrently (Merelman 1973):

1. The marks of status in a yeoman society, 1787–1857
2. Education as a preparation for life, 1857–1945
3. Education as a means of fulfilling human potential, 1945–

Given that Merelman published his typology in 1973, right after the last Apollo mission and when Internet was in its infancy as ARPANET with only about 40 connections, we can wonder if a fourth mode of adaptation is emerging today, somehow preparing students to become nodes in an information-network civilization. But his three categories do provide a framework for thinking about the current situation. Mode 1 emphasized the humanities as marks of cultivation among the children of the elite, in order to authenticate their elevated social status even though they had not yet accomplished anything in life. Mode 2 prepared students to fill well-defined roles in industrial society, as reliable workers in jobs that required fairly standard skills. Mode 3, which Merelman criticized but reported had become fashionable in the educational profession, was psychological in orientation, rather than cultural or technical like the earlier two. It postulated that each child should gain personal fulfillment in life, even without elite status or a respected job. From the standpoint of the spaceflight social movement, some synthesis of Mode 1 and Mode 2 would be most appropriate, producing technically sophisticated leaders. Mode 3, which promoted emotional adjustment, might be counter-productive for space exploration, yet appropriate for a world in which social stability is the main goal.

Setting education aside, we can consider how gender-related cultural values interact in a different way through the preference for science fiction novels in the 1981 University of Washington study, by correlating that variable with two statements about the space program. With the genders combined, 556 students disliked SF novels, rating them 0–2 on the 0–6 preference scale. Among them, 50.3 % agreed, "In the long run, discoveries made in our space program will have a big payoff for the average person." Among the 198 who gave the top 6 rating to SF novels, 81.2 % agreed. The tau-c correlation is 0.25. Now separate all the respondents by sex, 682 women and 741 men. Among women the correlation is 0.19, and among men, 0.26, slightly different but both positive and statistically significant. A negatively phrased item gives similar results: "The United States is spending too much money on space, so appropriations for the space program should be reduced." For all 1,439 students the tau-c was −0.24; for women it was −0.17, and for men, −0.25.

For both genders, science fiction was associated with increased support for space, slightly more so for men. Perhaps science fiction fans will always be a tiny minority of the population, but all people are increasingly living in a science fiction world, because of information technology. Much rhetoric trumpets the radical notion that today we are going through a general cultural revolution comparable to the Neolithic Revolution that gave humanity agriculture and cities, and the Industrial Revolution that produced increased prosperity and societal complexity. Perhaps, for once, overblown rhetoric is correct, and we are entering a new Information Society that will transform all aspects of life, in ways both small and large, setting the stage for entirely different motivations for spaceflight.

The winter of 2013–1014 has been harsh, cold and snowy, so waiting for the bus to commute from home to my work at the National Science Foundation has been unpleasant. Therefore, each morning I clicked the link on my computer's browser to access the WWW page that tells me when the next bus will arrive at my local stop, thanks to the space program. What I experience is entirely related to personal computing and electronic communications, but the way the system works is for a GPS unit on the bus to receive signals from Global Positioning System satellites, send its geolocation to the central database, from which an arrival estimate is sent out over Internet for each of its regular stops.

The connection in the public mind between spaceflight and the information society we all live in today varies, depending upon exactly which questions we ask. The previous chapter showed that many computer games depict spaceflight, yet that does not mean that gamers necessarily favor the real space program. In 2000–2004 the General Social Survey asked 1,127 respondents whether they had visited a website related to computer games they played in the past month and administered the space funding item to them. Among the 675 respondents who had not checked a game-related website, 17.2 % felt too little was being spent, compared with 17.9 % of the 452 who had done so, not a big difference. But when the question concerned actually playing games online, asked only in 2000 and 2002, there was a significant difference. Fully 24.2 % of 165 online gamers felt space program funding was too little, compared with 15.5 % on non-players. *Star Wars Galaxies* did not launch until 2003, so those 165 gamers were pioneers, exploring the new possibility of virtual worlds substituting for actual other planets.

In 2000, the GSS asked, "Do you personally ever use a computer at home, at work, or at some other location?" A majority, 57.8 %, answered yes. That majority continued to grow, reaching 66.8 % in 2002, 73.9 % in 2004, 79.8 % in 2010, and 80.5 % in 2012. In 2000, 18.9 % of the 656 who used computers felt space funding was too little, compared with 9.3 % of those who never used computers.

One might have expected the difference to diminish over the following years, as people who lacked a pioneer spirit began to use computers, but that is not what happened. In 2012, fully 27.9 % of computer users felt space funding is too little, compared with 11.9 % of those who do not use computers.

It is too early to come to firm conclusions, but here are some plausible theories. Yes, prosperous, educated, pioneering people were among the first to enter personal computing, and such people would be expected to be more enthusiastic about spaceflight than the average. But the Internet revolution may be even far more revolutionary that we realize, promoting among the general public radically new perceptions of science and technology across all fields, including spaceflight. This indirectly may effect everybody, even those who are not computer savvy. Thus, conceivably the apparent increase in support for the space program in the most recent years may result not from anything NASA has done, but from the dawn of the Information Society.

10.4 Freefall Is not Decline

In the *Flash Gordon* movie series, at one point the rockets cut out on Dr. Zharkov's spaceship, and he exclaimed, "We're falling through space!" He should take a remedial astronautics class, because the zero gravity experienced in real spaceflight is *freefall*, and it does not prevent a vehicle from reaching its destination. Perhaps the most optimistic result of the research reported in this book is what it did not find—any evidence that the general public has turned its back on spaceflight. The Space Race of half a century ago raised public awareness of many meanings that spaceflight can have, and that level of understanding persists. There is an obvious astronautics metaphor: The space program is coasting in low Earth orbit, having already accomplished amazing things, and its current freefall need not be a warning of disaster. If this is a good metaphor, rising to a higher orbit and eventual escape requires less delta-V (velocity increase) and can be done gradually.

We do need more social scientific research. This book has intentionally restrained the impulse to indulge in complex statistical analysis that might disentangle cause from effect from spurious correlations. Unfortunately, few social scientists have shown much interest in poll data about the space program, but that sad situation may be a boon for a few motivated students who can complete high-quality doctoral dissertations, delving deeply into the data we have surveyed here, testing some of our hypotheses, and especially looking into connections between spaceflight and the new information technologies. The fact that data from the Harris polls and the General Social Survey are available free online would not only facilitate their work, but illustrates the radical implications of Internet. It will also be essential for well-funded research studies conducted by experienced social scientists to collect new data on the ideas of scientists, opinion leaders, and other unusually influential groups.

As used in politics and commerce, public opinion polls are a tool for persuasion, measuring the popular mind in order to manipulate it. Standard social science theories recognize the disproportionate influence of opinion leaders and other elites in shaping a society's culture, and in motivating collective action. Given that the public has been lead into spaceflight by activists, we can wonder what will happen if the new information and communication technologies strengthen democracy, thereby reducing the power of leaders. If spaceflight really can improve life on Earth, as well as opening up new possibilities for human action beyond its atmosphere, space enthusiasts will have nothing to worry about.

References

Astor, J. J. (1894). *A journey in other worlds* (p. iii). New York: D. Appleton.

Bainbridge, W. S. (1982). The impact of science fiction on attitudes toward technology. In E. M. Emme (Ed.), *Science fiction and space futures* (pp. 121–135). San Diego: American Astronautical Society.

Bainbridge, W. S. (1986a). *Dimensions of science fiction* (p. 171). Cambridge, Massachusetts: Harvard University Press.

Bainbridge, W. S. (1986b). *Dimensions of science fiction* (p. 168). Cambridge, Massachusetts: Harvard University Press.

Bainbridge, W. S. (2007a). *Across the secular abyss*. Lanham, Maryland: Lexington.

Bainbridge, W. S. (2007b). *Nanoconvergence*. Upper Saddle River, New Jersey: Prentice-Hall.

Bainbridge, W. S. (2013). *eGods: Faith versus fantasy in computer gaming* (pp. 3–4). New York: Oxford University Press.

Bush, V. (1980). *Science, the endless frontier*. New York: Arno.

Drexler, K. E. (1992). *Nanosystems: Molecular machinery, manufacturing, and computation*. New York: Wiley.

Gibbon, E. (1880). *The decline and fall of the Roman Empire*. New York: Hurst.

Huntington, S. P. (1996). *The clash of civilizations and the remaking of world order*. New York: Simon and Schuster.

Launius, R. D. (2010). Can we colonize the solar system? Human biology and survival in the extreme space environment. *Endeavour, 34*(3), 122–129.

Merelman, R. M. (1973). Public education and social structure: Three modes of adjustment. *The Journal of Politics, 35*, 798–829.

Nathan Keyfitz, N. (1987). The family that does not reproduce itself. In K. Davis, M. Bernstam, & R. Ricardo-Campbell (Eds.), *Below replacement fertility in industrial societies* (pp. 139–154). Cambridge, England: Cambridge University Press.

National Center for Science and Engineering Statistics. (2013). *Women, minorities, and persons with disabilities in science and engineering: 2013* (p. 4). Arlington, Virginia: National Science Foundation.

National Science Foundation. (1999). *Women, minorities, and persons with disabilities in science and engineering: 1998*. Arlington, Virginia: National Science Foundation.

Niebuhr, H. R. (1929). *The social sources of denominationalism*. New York: Holt.

Ogburn, W. F. (1922). *Social change*. New York: Huebsch.

Plutzer, E., & Berkman, M. (2005). The graying of America and support for funding the nation's schools. *The Public Opinion Quarterly, 69*, 66–86.

Roco, M. C., & Bainbridge, W. S. (Eds.). (2001). *Societal implications of nanoscience and nanotechnology*. Dordrecht, Netherlands: Kluwer.

Roco, M. C., & Bainbridge, W. S. (Eds.). (2003). *Converging technologies for improving human performance*. Dordrecht, Netherlands: Kluwer.
Roco, M. C., Bainbridge, W., Tonn, B., & Whitesides, G. (Eds.). (2013). *Convergence of knowledge, technology and society*. Berlin: Springer.
Snow, C. P. (1959). *The two cultures and the scientific revolution*. New York: Cambridge University Press.
Sorokin, P. A. (1937–1941). *Social and cultural dynamics*. New York: American Book Company.
Spengler, O. (1926–1928). *The decline of the west*. New York: Knopf.
Toynbee, A. J. (1947). *A study of history*. New York: Oxford University Press.
Turner, F. J. (1920). *The frontier in American history*. New York: Holt.
Wagner, R. (1994). *The artwork of the future and other works*. Lincoln, Nebraska: University of Nebraska Press.
Wilson, E. O. (1998). *Consilience: The unity of knowledge*. New York: Knopf.
Zachary, G. P. (1997). *Endless frontier: Vannevar Bush, engineer of the American Century*. New York: Free Press.

Index

W.S. Bainbridge, , *The Meaning and Value of Spaceflight*,
Space and Society, DOI 10.1007/978-3-319-07878-6

MIX
Papier aus verantwortungsvollen Quellen
Paper from responsible sources
FSC® C105338

If you have any concerns about our products,
you can contact us on
ProductSafety@springernature.com

In case Publisher is established outside the EU,
the EU authorized representative is:
Springer Nature Customer Service Center GmbH
Europaplatz 3, 69115 Heidelberg, Germany

Printed by Libri Plureos GmbH
in Hamburg, Germany